THE BALANCE OF IMPROBABILITIES

A benign view of the Regius Professor of Medicine at Oxford. From a painting by Alexandra Harris, 1985.

THE BALANCE OF IMPROBABILITIES
A SCIENTIFIC LIFE

HENRY HARRIS

OXFORD UNIVERSITY PRESS
1987

Oxford University Press, Walton Street, Oxford OX2 6DP
Oxford New York Toronto
Delhi Bombay Calcutta Madras Karachi
Petaling Jaya Singapore Hong Kong Tokyo
Nairobi Dar es Salaam Cape Town
Melbourne Auckland
and associated companies in
Beirut Berlin Ibadan Nicosia

Oxford is a trade mark of Oxford University Press

Published in the United States
by Oxford University Press, New York

© Henry Harris, 1987

All rights reserved. No part of this publication may be reproduced, stored in a retrieval system, or transmitted, in any form or by any means, electronic, mechanical, photocopying, recording, or otherwise, without the prior permission of Oxford University Press

British Library Cataloguing in Publication Data
Harris, Henry
The balance of improbabilities:
a scientific life.
1. Harris, Henry.   2. Medical scientists —
Great Britain — Biography.   3. Cytologists
— Great Britain — Biography.
I. Title
610'.92'4   R489.H2/
ISBN 0-19-858217-X

Library of Congress Cataloging in Publication Data
Harris, Henry, 1925–
The balance of improbabilities.
Includes index.
1. Harris, Henry, 1925–   .   2. Cytologists —
England — Biography.   3. Physicians — England — Biography.
I. Title.
QH31.H34A3   1987   574.87'092'4[B]   86-23530
ISBN 0-19-858217-X

Phototypeset by Dobbie Typesetting Service, Plymouth, Devon
Printed in Great Britain by St. Edmundsbury Press
Bury St Edmunds, Suffolk

For Paul, Helen, and Ann

PREFACE

What follows is an account of my life as a scientist. My private life, the family I grew up in and the one that grew up around me, receive scant mention in these pages. But what I have written is not intended only for scientists. My hope is that the intrinsic interest of an overtly unruffled scientific life, as mine has been, can be made comprehensible even to those for whom science remains an unfamiliar world. One obvious difficulty is, of course, the translation of scientific ideas, without falsification or trivialization, into the language of everyday life; but it is even more difficult to capture in words the high tension of a scientist's inner life. The public life of most scientists is very dull, but their obsessional struggle to wrest new knowledge from an unyielding world is not. Any scientific investigation that is more than a series of routine measurements has within it the essential elements of a good detective story; and it is a pity that so few scientific biographies or autobiographies have the intensity of thrillers. I do not imagine that I have produced a thriller, but I have done what I can to convey something of the excitement that can induce a man to devote his entire life to the pursuit of science, whatever the outcome of his endeavours and whatever the cost. If this sounds a little too dramatic, let me hasten to add that doing experiments has also been great fun; and so has the writing of this book.

Oxford
Hilary, 1986   H. H.

CONTENTS

1.	A schooling in Australia	1
2.	*Sidere mens eadem mutato*	23
3.	Interlude in a southern city	48
4.	The young Australian at Oxford	54
5.	On my own	84
6.	The USA	117
7.	The green world	138
8.	Return to Oxford	163
9.	Cell fusion	185
10.	Cancer	207
11.	The Queen's Professor	224
	Index	235

1
A SCHOOLING IN AUSTRALIA

My nostalgia begins with blurred images of Sydney sweltering under an Australian summer sun. First, Bronte Beach in 1930, where Sam Harris, a penniless immigrant from an obscure country town in the Russian Pale of Jewish Settlement, his wife, and their two sons occupy two rooms over a milk-bar close to the crowded water's edge. Then a dilapidated weatherboard house with a corrugated iron roof in the fag-end of Lidcombe, itself at that time a fag-end suburb in the infinitely monotonous hinterland of the city. Then a small flat in a strictly rectangular red-brick building on the edge of middle-class Kensington. And, finally, in sharp focus, a semi-detached bungalow, No. 8, Avoca Street, Bondi. This rapid ascent from penury to modest suburban security had been propelled by the energy and ingenuity of my father. His solution to unemployment in the depth of the Great Depression was to open a small confectionery business right in the commercial centre of the city, entirely on credit. Miraculously the business survived, and it remained our mainstay until my father retired. All that is left of it now is a small pile of empty chocolate cartons, bright red and shiny, with Harris Bros in bold black cursive on the sides.

My formal education started at Auburn Public School, an undistinguished example of the New South Wales State primary school system, but with the solid advantage that it was within walking distance of our house in Lidcombe. My brother and I, elegantly turned out in new trousers that reached well below the knee (to allow for growth), arrived there one morning and were at once initiated into the barbaric rituals of juvenile self-government. I cannot recall the classroom or the face of my teacher, but the small asphalt playground comes vividly to mind. Our survival depended on the outcome of a series of gladiatorial combats staged there during the lunch hour. My brother was a heavy boy and must have given a good account of himself, for the baiting soon stopped. I think I managed to hold my own; but there is a sense in which it is true to say that I have been fighting ever since.

Auburn Public School gave way a few months later to Kensington Public School, where I was admitted to Class 1A. My recollection of the children there is that they were less aggressive than those I had encountered at Auburn, but this may simply be a reflection of the fact that I had by this time learned to pass unnoticed. My clearest memory of the school is again the playground: it contained a large peppercorn tree whose fruit crushed between the fingers or underfoot provided a rich source of aromatic pleasure. I learned to play games, to swop cigarette cards, and to perform creditably in the little tests that monitored our progress; and I soon became a major contributor to the juvenile street life that filled the hours between school's end and the evening meal. But we did not stay long in Kensington either, and it is Bondi that provides the first detailed vision that I have of my early life. The semi-detached bungalow at No. 8 Avoca Street was to be our home for some four years, and it was there that I first threw roots.

Avoca Street was an ill-assorted straggle of ageing terrace houses patched at one end by a small group of identical bungalows of which No. 8 was one. The street sloped away from a small suburban shopping centre in the Bondi Road along which trams from the city clattered down to Bondi Beach a couple of miles further along. At the junction of Avoca Street and Bondi Road, there stood, on one side, a cinema, Hoyt's Olympic No. 2, and, on the other, a grocer's shop where in due course I learned that a penny could sometimes buy a remarkably large volume of broken biscuits. In Wellington Street, on the opposite side of the Bondi Road, there loomed a large late-Victorian church and, over the road, the rather forbidding edifice of Bondi Public School where I was destined to complete my primary education.

It was there that an image of England first appeared on my conceptual horizon. Australia in the 1930s was still very close to being a British colony and, although public assertions of cultural and political independence were constantly being heard, the patterns of daily life and the attitudes adopted subconsciously by most of the population were unmistakably British derivatives. In the summer heat, which in Sydney could, at times, be as fierce and as humid as in Singapore, business men wore sober three-piece suits to work, looking not very different from London bank clerks. At Christmas, with the temperature perhaps '100 degrees in the shade', the shop windows displayed Santa Claus in snow scenes, and the traditional

Christmas dinners ground to a halt with the traditional Christmas pudding. The most important social event of the year was the State Governor's annual afternoon tea party, and the receipt of an invitation to this conferred a social cachet not attainable by any other means. At Bondi Public School, morning assembly ended with the pupils springing to attention and declaring in unison: 'I honour my God; I serve the King; and I salute the Flag'. And salute it we did. The unofficial national anthem was Peter Dodds McCormick's 'Advance Australia Fair', which we sang lustily at frequent intervals. Its second verse, now suppressed, runs thus:

> When gallant Cook from Albion sail'd
> To trace wide oceans o'er,
> True British courage bore him on
> Till he landed on our shore;
> Then there he raised Old England's Flag,
> The standard of the brave;
> 'With all her faults we love her still'
> 'Britannia rules the waves'.

Our schooling introduced us to a remarkably disembodied world. Self-conscious Australian historiography was only just beginning, and our lessons in Australian history were limited to romanticized versions of the colonization of the Continent and of the great overland explorations. The natural focus of any deeper historical sense that we might have had was England. I expect we were as familiar as English schoolboys would have been with Magna Carta and the Reform Bills, with Stephenson's Rocket and Arkwright's Spinning Jenny. We were committed participants in the great battles of English history, and we gazed with admiration at the British Empire splashed in pink across the globe. Of the history of the rest of the world, we were taught and knew nothing. Literature, as we studied it, was almost entirely the literature of England, selected in accordance with the most conventional canons of Edwardian taste. It is difficult for those who have not been exposed to it to appreciate how dislocating an experience it is to be constantly immersed in an imaginary world that in no way reflects the physical environment in which one spends one's days. The poetry we read at school dealt with skylarks, nightingales, white hawthorn, yellow leaves driven by the West Wind, images that became realities for me only twenty years later. The London of Dickens and the gentle English countryside of Jane

Austen found little echo in Avoca Street, Bondi. All we knew of Australian literature, apart from an occasional trite poem by Henry Kendall or Dorothea Mackellar, which we were made to memorize, were the short stories of Henry Lawson and the bush ballads, and these, to a city boy, were no less the products of an alien culture. But I have tangible evidence that the bush ballads must have made a strong impression. The school prizes given at the end of each academic year were invariably books and, within the limits of what was judged appropriate, the recipients were permitted to make their own choices. My prize for the third class (mainly seven- or eight-year olds) was an edition of the *Collected Poems of A. B. (Banjo) Patterson*, king of the bush balladists.

A similar relic from the fourth class gives evidence of another cultural influence which, I suspect, was at that time even more powerful than school. On my shelves there still sits a fat green volume into which an English translation of both parts of *The Count of Monte Cristo* has been squeezed. This is no ordinary book. It is the book of the film; and for a frontispiece it has a colour photograph of Elissa Landi (Mercedes) with whom I was then desperately in love. Every Saturday afternoon my brother and I spent at the Hoyt's Olympic No. 2. This was our window on the world. Between 1932 and 1935, when we moved away from Avoca Street, I saw virtually every English and American film considered in Australia to be suitable for Saturday matinee viewing. The entry fee to this recurrent feast of the mind was sixpence. The films I saw then had profound, even if transitory, effects. After seeing *The Invisible Man* I spent hours in our small back garden (back yard, as it is known in Australia) tipping coloured fluids from one bottle into another in a frenzied effort to create the magic potion for myself. After *The Count of Monte Cristo* my brother and I fought duels with each other for weeks. And I was constantly in love with one beautiful woman after another, Dolores del Rio, Myrna Loy, Merle Oberon, Loretta Young, but, above all, Elissa Landi who, in my eyes, embodied in supreme degree all womanly virtues.

My passage through the first years of primary school was uneventful and crowned annually with a General Proficiency prize. Girls were separated from boys at the end of the second class, at the age of six or seven, and they were not reunited until they entered university. The ethos of the male half of the Bondi Public School was very male indeed. I fought Archie Cocksedge and Johnny Morgan, both of whom I liked, simply to establish the pecking order; and I fought a less likeable colleague because he had stolen some of my marbles. With Vince Helby and Hector Wilson I competed for the honour of being top of the class, a distinction far less highly regarded than athletic prowess or martial vigour, but important to me none the less. A modest skill in the water earned me the Bronze Medallion of the Royal Life Saving Society but, being myopic from an early age, I was to my chagrin a less than indifferent cricketer. I do not know what has become of any of those friends of days gone by, but I hope that life has treated them gently.

There was one private extracurricular activity about which I said nothing at school. My parents retained an ancestral tradition of encouraging musical ability in their children, even when there was none. Both my brother and I were subjected to piano lessons given once a week in our home by a Miss Harcourt, a rather overweight, but still glamorous, creature, as incongruous in that environment as a polar bear in the tropics. We had acquired a second-hand upright piano, of a make that I have never seen since, and with this instrument I did daily battle for several years. The weekly visits of Miss Harcourt were eventually replaced by weekly trips to the New South Wales Conservatorium of Music, where Miss Harcourt's duties were taken over, at a more advanced level, by a Mr Ramsey Pennicuick. I never did become a competent musician, but it was not all wasted effort, for what remained has, from time to time, given me great pleasure.

It was during our Avoca Street period that I received my definitive initiation into the Australian beach life. Bondi Beach was a half hour walk from our home or, if time pressed, a ten-minute ride in a rackety tram that ground down the hill to the waterfront. There appears to have been a clear moral obligation to go surfing every weekend throughout the summer and after school on weekdays as well when the hot weather became oppressive. I became a moderately competent surfer, but I must confess that this occupation never did fill me with the enthusiasm that it appeared to generate in my school friends, or, for that matter in my father, who remained a devotee of the surf to the end of his life. I quickly became bored with the ritual of sand and sea and, as I grew older, the beach came to occupy less and less of my leisure hours. It eventually became no more than a place I would visit once in a while to cool off in the hot weather.

If the beach life did not enchant me, the life of the streets certainly did. As soon as my school books were dumped and my mother informed of any notable events that had occurred in the course of the day, I was out on the streets, alone or with my brother, looking for adventure and sometimes for trouble. Avoca Street at its tail-end led into a maze of similar streets, but these soon petered out into sandhills and a scruffy 'glen' that meandered down to the then almost entirely undeveloped Tamarama Beach. In the late afternoon this area of decadent bush was the haunt of various juvenile gangs that sought, with only modest success, to terrorize each other. My brother and I had occasional brushes with these raiding parties, but always seemed to come away unscathed. We instinctively took on the appropriate protective coloration, became masters of working-class idiom, which distinguished friend from foe, and were probably unexcelled in the competitive boasting that seemed then, and perhaps still is, the normal mode of communication of young adventurers in the streets. As I look back on it now, it is clear to me that what I found in those streets was the real Australia—urban, radical, aggressive, and swayed by deep, if narrow, loyalties. Imperceptibly, but inevitably, some of these attitudes and the myths that generated them seeped into my own view of the world; and I am not sure that I have ever completely shed them.

Harris Brothers prospered and, as I entered my final year at Bondi Public School, my father decided that we were affluent enough to buy a home of our own. 15 Poate Road, Paddington was a nondescript red-brick bungalow built some time after the First World War, and there was nothing about its style or construction that distinguished it in any important way from the endless rows of similar houses that made up all but a small part of the city of Sydney. It was, however, perched at the top of a sloping site, so that from the front gate you looked up at the house across a long stretch of garden, and a flight of tiled steps flanked by two tall cabbage palms led up to the front door. Overlooking the street in this way, the house had a rather superior air, and when I first set eyes on it I could scarcely believe that we could be the owners of something so grand. Its outer walls formed an unbroken rectangle with the front door at the mid-point of one wall and the backdoor, complete with flyscreen, directly opposite. Within, the standard dispensation: three bedrooms, living room, dining room, breakfast room communicating with the kitchen, and bathroom; to which were added, in the best Australian tradition, a laundry and a lavatory accessible only from the back garden. The garden contained a few mature fruit trees, including a fig and a loquat, and a shed against which a trellis-work arbour had been constructed. I was destined to spend a good deal of my spare time in that arbour, for it was a cool place screened from the sun by an overgrown grape-vine. Adjoining the house on one side was a scruffy field that had not yet been developed; and for some years this formed a natural extension of the amenities available to us. It was there that I met and made fleeting friendships with the young inhabitants of the houses round about. No. 15 Poate Road remained the family home until my brother and I left it to make our separate ways in the world.

I got to know the streets of Paddington as well as I had known the streets of Bondi. Paddington offered greater contrasts. Close to Centennial Park, and especially in Lang Road which ran beside it, there were some very grand houses set in large gardens. It was impossible to walk along Lang Road without giving way to vivid fantasies about the lives of the people who could afford to live on such a scale. As you moved away from the park, the houses rapidly became conventional Australian middle-class, and these in turn gave way to the Victorian terraces that lined the maze of narrow streets on both sides of Oxford Street. These terrace houses, now renovated and fashionable, were then in a state of incipient decay and were inhabited by a sub-population whose reputation was such as to deter entry into those areas at night. I was unable to make any contact with the children in the Lang Road houses, but struck up a number of interesting friendships with youngsters from the terraces. As time went by, however, it became clear to them that my aspirations were very different from their own, and this led in due course to my being rejected and finally resented. None of these street friendships survived my entry into university.

Oxford Street, lined with small shops, was a delight especially reserved for Saturday mornings. Greengrocers in Sydney were then almost entirely immigrants from Sicily or the toe of Italy; the fishmongers were Greeks. Fruit and vegetables came with a flow of barely comprehensible Anglo–Italian repartee interspersed with snatches of Puccini; fish (and chips) came wrapped in newspaper. There was no such thing as a delicatessen; the very word was unknown in Sydney at that time. But there was an establishment known as a ham-and-beef shop where, among other things, you could indeed buy ham and corned beef. The dairy produce shop was run by an impeccably neat Esthonian family, and there you could buy eggs said to be fresh from the farm and two, but only two, kinds of cheeses: mild and tasty. The newsagent sold everything that wasn't edible. A small private lending library, where twopence gave you possession of any book for a fortnight, was the source of my exhaustive knowledge of the life and works of Bulldog Drummond. A restaurant called 'The Ritz' offered a three-course meal for one shilling and threepence. The son of the owner was one of my street friends but, though we were together a great deal, I never did get a free meal at The Ritz. And then there was Centennial Park itself. I spent so much time there that parts of it became as familiar to me as my own garden. Once, after a rebellious scene with my parents, I spent a night in a hollow at the base of a huge Moreton Bay fig tree which gave shade to one of my favourite haunts.

The move to Poate Road introduced a further, unforeseen, element of adventure into my daily life. As it had been decided that I should remain at Bondi Public School until I had taken the Qualifying Certificate (QC) Examination for entry into the State secondary school system, I made the daily journey from Paddington to Bondi and back by tram. Those old trams were magnificent. Each tram-car contained eight compartments, two at each end for smokers, four in the middle for non-smokers. The non-smoking compartments were enclosed by wooden folding doors with glass panels; the smokers had brown canvas roller blinds that were pulled down only when it rained. Naturally I chose to travel in the smoking compartments where the men were, the inner compartments being dominated by women. The journey to school took about twenty minutes during which (except when it rained) I had an unimpeded view of all life going by. The tram skirted Centennial Park, dawdled past the long line of small shops into Bondi Junction, laboured up the hill to Waverley Park, and then tore down the Bondi Road to Wellington Street. No two journeys were the same, and no day passed without some observation of interest and importance. As the trams going out to the suburbs in the morning and back into the city in the afternoon were usually uncrowded, I had a free choice of where I sat and could thus adjust the passing scene to suit my mood. It was my first taste of power and independence.

For those at the head of the class, the final year at primary school was a period of intensive coaching for the QC Examination. At the pinnacle of the State secondary school system stood the Sydney Boys' High School, to which entry was highly competitive and decided entirely by the results obtained in that examination. All over the city, the brightest boys were trained for the event, and the reputation of a teacher rested to some extent on the success of his pupils in gaining access to Sydney High. There were three of us in my class who were thought to have a chance, and we were given extra tuition on Sunday mornings at the home of our teacher, an energetic man named Townsend who, under the influence of our lessons in English history, was given the nickname 'Turnip'. He worked hard at his job, and all three of us made it. The rejoicing in my family knew no bounds, and I at once became the epicentre of my parents' ambitions. My brother had previously gained access to the Sydney Boys' Technical High School, for at that time the idea was that he should become an engineer; but he was never as expert as I was at winning school prizes, and school prizes were the coinage that my parents understood best.

The Sydney Boys' High School, known throughout the city as the High, was and still is a low grey rectangle built around a courtyard and set back from the road (Anzac Parade) in extensive grounds. The front façade, which looked out over playing fields, was dramatized by a row of Doric columns, and the surrounding parkland, containing large trees and flowerbeds, seemed after the asphalt patch of Bondi Public School a glimpse of paradise. The Boys' School was separated by a high but penetrable barrier from the Sydney Girls' High School, the corresponding pinnacle of the female side of the secondary school system; but, as befitted the Australia of those days, neither the buildings nor the grounds of the Girls' School were as impressive as their male counterparts. From Poate Road I walked to school down Moore Park Road beside the high brick wall that surrounded the Royal Agricultural Showground, diagonally across the cricket fields of Moore Park, and then along the wide tree-lined avenue of Anzac Parade down which the trams rattled towards Maroubra and La Perouse. When the weather was bad I took a double-decker bus from Poate Road to the bottom of Moore Park Road, and from the upper deck I could see over the brick wall into the Royal Agricultural Showground. There was often something of interest there, but it was a poor substitute for the tram ride to Bondi.

Unlike my brother, the prospective engineer, I had not yet formed any clear idea of what I proposed to do with myself. In the first year at the High there was no choice in the curriculum. We all did maths, which included an introduction to algebra, English, history (a thumbnail sketch of the ancient world) and elementary science, and were introduced to Latin and French. Nearly all the masters bore nicknames that encapsulated some peculiarity of physique or behaviour ('Gummy' Evans, 'Fizz' Hanly, 'Jerker' Jerrems) and from the word go they set about training us for academic success. The competition was very fierce. Because of the highly selective mode of entry, the school contained an inordinately high proportion of gifted boys and, whereas scholastic ascendancy had cost me little effort at Bondi Public School, I soon found that at the High the going was much tougher. We were given a test at the end of each term and a formal examination at the end of the year. My parents were aghast to learn from the written report they received at the end of my first term that I had slipped to being seventeenth in a class of forty-two; and there was deep concern in my mother's voice as she asked me whether I had lost my touch. Those early termly reports were remarkably astringent. I remember one that informed my parents that I was much too careless ever to become a competent linguist, and another that doubted my ability ever to become a scientist. But all was not lost. Much to my parents' relief, my more concentrated efforts were crowned at the end of the year by another General Proficiency prize.

It was during my first year at the High that I became interested in science. This interest was sparked off by a series of romanticized sketches called *Living Biographies of Great Scientists* that I had received as a gift. I had also been given a companion volume, *Living Biographies of Great Composers*, but found that less interesting. Paul de Kruif's popular accounts of the great bacteriological discoveries and a series of similar books followed. There is only one of these many potted biographies that I still remember in any detail and that, curiously enough, is an account of the life of Wöhler. Wöhler is not a household name and is now remembered in textbooks of the history of science mainly for the synthesis of urea, the first constituent of the body to be made in the laboratory from simple inorganic substances. This struck me then, and still does, as a marvellous achievement, not only as one of the foundation stones of organic chemistry, but more importantly because it demolished at one blow long-held assumptions about the nature of life. After Wöhler, those who thought about these matters at all had to come to terms with the fact that living forms were part of a larger unity from which they were separated by no clear demarcation.

I have always reserved my greatest admiration for discoveries that overturn cherished assumptions, discoveries that make us see the world differently. This is not simply iconoclasm, but a considered view of what constitutes originality in science. It must have been an unconsidered view that I held very early on. But there was something else about Wöhler's discovery that had a special appeal for me. My father was an uncompromising atheist of an old-fashioned, anti-clerical type, to whom all religious beliefs were mere superstition. Only social obligations imposed by marriages or deaths could induce him to attend a place of worship and he was always ill at ease there. Religion, like politics, is, at least initially, largely inherited, and I adopted my father's views as a matter of course. There was never any soul-searching. I simply preferred the clear light of day to the chiaroscuro of mysticism. Scientific investigations that reduced obscurity, that demystified, that provided natural explanations for supernatural ones fitted in with my view of the world and lent support to it. I think the special appeal of that popular account of Wöhler's work must have had something to do with the element of demystification in it. The passage of the years has tempered but not fundamentally modified my outlook, and discoveries that dissipate mystery still hold my admiration.

A book that made a serious impression at that time was Sherwood Taylor's *The World of Science* which I received as a gift from a Mr Abraham Goldstein, whose shop near the Bondi Junction announced to a sceptical world that he was a Genuine Swiss Watchmaker. Despite years of residence in Australia, Mr Goldstein (I never heard him called anything else) had acquired little English, so little indeed that his incomprehension frequently generated grotesque situations that became legendary. He was, for example, enormously impressed on one occasion by a sign in the window of a neighbouring dry-cleaning business and ordered a similar sign for his own shop. Despite the protests of the sign-writer, Mr Goldstein insisted that the sign should be exactly the same as the one he had seen, with the result that for several months his shop window was graced by a splendid sign announcing that all work done was guaranteed, but that no responsibility was taken for fading. He had an irrational fear of having stolen watches foisted on him and precipitated numerous crises by attempting to cross-question his customers in a totally incomprehensible linguistic mishmash about the origin of their timepieces. He was thus a universal figure of fun, but there was, beneath all the drollery, a fundamental dignity that made it impossible for anyone to treat him with disrespect. He was a crony of my grandfather's and once in a way would pay a visit to our home. His visits were always stylish performances. He came dressed impeccably in a pin-striped dark-blue suit with razor-sharp creases in his trousers and some flower, usually a carnation, in his buttonhole. He always brought a huge bunch of flowers for my mother and small gifts of one kind or another for my brother and myself. At table he ate elegantly and sparingly, unlike most of our visitors, spoke little himself, and paid close attention to what others said. He never outstayed his welcome and made a ceremonial departure, showering gracious compliments on my mother.

It might seem rather odd that Mr Goldstein should have been the donor of my long-cherished copy of *The World of Science*. It came about in this way. My grandfather, as was his habit, had informed him of some success I had achieved at school, and Mr Goldstein, beneath the genuine Swiss exterior a traditional Hasidic Jew, at once decided that he would show his respect for learning by buying me an expensive book. He asked my parents for permission to take me out to afternoon tea and make the purchase. On the appointed day he arrived, impeccably dressed as usual, and took me into the city where we had afternoon tea together at Repin's Coffee shop in Market Street. Mr Goldstein regaled me with waffles, which he had some difficulty in ordering because the waitress did not at once understand what 'vufflies' were. Replete, we then went round to Angus and Robertson's bookshop in Castlereagh Street, and I was given the liberty to choose any book I wanted. *The World of Science* was the choice I made. I do not think it hopelessly romantic to suggest that the direction that my life ultimately took might perhaps have been influenced just a little by what Mr Goldstein did that day.

If I had been pressed at the age of eleven to say what I really wanted to become, I think I would have admitted, without too much self-consciousness, that the idea of being a famous bacteriologist seemed very attractive. But science was soon dislodged from my mind by an extraordinary event that had little to do with the direction of my academic talents. At school, some importance was attached, although not by the boys themselves, to public speaking. There was a special prize for oratory donated by some clergyman about whom nobody seemed to know anything. The competition for this prize was divided into two sections, one for the junior and one for the senior school; and the two awards were made on the strength of histrionic performances offered by those who felt confident enough to air their views on a subject of their own choice for a period not exceeding ten minutes. In the speech I made, I presented a definitive solution to the problem of Universal Peace and thereby became the first boy in the history of the school (or so I was told) to win the junior prize in his first year. My General Proficiency prize was thus enhanced by a chrome-plated trophy on which my oratorical prowess was engraved. I won this oratory prize in the appropriate division every year that I was at school, and in my last two years became the captain of the school debating team. We competed with teams from other secondary schools for the Hume-Barbour Trophy, a splendid bronze object bearing a statue of Demosthenes. Two small bronze plaques which I still possess testify to the fact that in each of the two years of my captaincy, The Sydney Boys' High School Debating Team was victorious. The precocious revelation of my rhetorical powers gave my father food for thought, and he began to envisage for his younger son a glittering career in the law. For the son it didn't much matter what the career was so long as it was glittering; but the idea of the law did begin to compete with the attractions of bacteriology.

There was, however, no need yet to take a decision. The school curriculum narrowed only very gradually, and irreversible specialization did not take place until you approached the final year. With one exception. At the end of the first year the more promising linguists were given the choice of taking a second language, either German or Ancient Greek, but not both, instead of history. I chose German because it seemed then that an additional living language might be of more practical use, with the result that I never again received a single history lesson and did not acquire even an elementary working knowledge of Greek. With this one barbaric variation, the range of subjects taken by all boys remained unchanged until after the Intermediate Certificate Examination which was taken in the last term of the third year and marked the end of the junior school. Having had a lifetime's experience with the lop-sided products of early specialization, I look back with some gratitude to that inflexible curriculum.

The secondary schools that offered a full five-year curriculum leading to university entrance were divided for purposes of competitive sport into two hostile groups. The State schools, which were free, formed an association known as The Combined High Schools. The old-established fee-paying schools, based on English models, styled themselves The Great Public Schools. The two organizations did not interact in any way, the Combined High School teams playing their games on Wednesday afternoons, the Great Public School teams on Saturday mornings. In this schismatic structure the position of Sydney High School was unique. It was, being a free State school, a member of the Combined High Schools *ex officio* ; but, being older than all but one of the Great Public Schools, it was also admitted into that organization, with the marvellous consequence that its pupils had a double dose of competitive sport—Saturday mornings as well as Wednesday afternoons. Sport thus loomed very large in the life of the school, especially in the minds of the younger boys for, as elsewhere in Australia, there was no glory like the glory of a great athlete. None the less, in the highly competitive academic atmosphere of the High, I do not think that preoccupation with sport ever reached quite the dimensions that one commonly found elsewhere. I discovered in my second year that my legs, although short, could move very quickly, and there was a period when it almost looked as if I might become a useful hurdler. But I was soon mowed down by the real athletes and relegated to the role of a committed spectator.

The school had a well-equipped gymnasium, and each week we had a compulsory class there supervised by a very military gymnastics master. Some of the boys reached an almost professional standard on parallel bars and vaulting horses, and the best of them gave public exhibitions of their skill. The gymnasium was also often used after school for boxing matches. We were forbidden to fight in the school grounds but were actually encouraged to settle our disputes in the gymnasium in accordance with Queensberry rules. When I first entered the school there was no school uniform, but a very conservative head master, whose signature appears in my prize books as Jas. A. Killip, succeeded in introducing one when I was half-way through. It took the form of a severe dark-grey suit with an embroidered version of the school's coat of arms applied to the left breast pocket. The introduction of a uniform provoked a good deal of resentment in the school and may well not have been accepted if Mr Killip had not somehow obtained funds to help boys for whom the purchase of a new suit would have been a hardship. I think the object of the exercise was that the High should conform more closely to the pattern of the fee-paying Great Public Schools, each of which naturally had an easily identifiable uniform. However, since access to the Great Public Schools was based on wealth, whereas access to the High was based on intellectual ability, the High remained, despite outward similarities, a very different place.

My early school years coincided with Hitler's assumption of power and the menacing crescendo of his Chancellorship. For what seems in retrospect to have been a remarkably long time, events in Germany were of no great moment to the man in the street in Australia. The first trickle of refugees from Germany and Austria began to arrive in Sydney, and my parents, who knew from their own experience where the shoe pinched, did what they could to help these often disorientated newcomers with their problems. They needed help even when they didn't need money, for the general reaction to these exotic strangers was one of hostility and extreme suspicion. Australians before the War were a remarkably insular lot and any immigration was instinctively felt to be a subversive threat. The Central European Jew, with his heavily accented speech, his curious clothes and above all his brief-case, was simply too much for the average Australian. The brief-case had a special significance. Before this immigration from Central Europe, brief-cases did not exist in Australia. Schoolboys, once they had abandoned the satchels that hung from the shoulder or were strapped on to the back, carried their books to school in small rectangular suitcases, usually made of a tough fibrous material called Cordite; and their elders used more expensive versions of the same sort of thing. Since the refugees were commonly referred to as 'refos', the brief-case became the 'refo-bag' and it was the mark of the beast. (Plate 1).

I do not think the children of the refos met with such hostility at the High. They came in small numbers into the junior school and were assimilated without too much difficulty. They were initially sought out as objects of curiosity and perhaps suspicion, but I cannot recall that they were ever systematically baited. For me they had a special interest because they brought a whiff of an extraordinary world which, through my increasing familiarity with the German language, had begun to take hold of my imagination. One of these boys, a powerful mathematician at school and later a philosopher, became a life-long friend. However, the most exotic creatures to join the school in my time came not from Central Europe but from

Plate 1. Walking along Anzac Parade to the Sydney Boys' High School: the Cordite suitcase, standard equipment before the arrival of the brief-case from Central Europe, and the double-decker London bus in the background.

England. One morning some of us were amazed to see arriving in their full glory two boys dressed in stylish morning suits, with grey striped trousers, black jackets, and shirts with large white collars. (I was later told, but could scarcely believe, that this really was the uniform of the school they had attended in London.) The head master sent them home at once to change. Just as well, for I am sure they would otherwise have been slaughtered. German Jews might be treated with suspicion, but upper-class Englishmen were beyond bearing.

As I moved through the junior school, preoccupation with events in Europe became more general. Some of the teachers began to comment on the day's news. A few had been Anzacs, either in France or in the Middle East, and they would occasionally give way to reminiscences. Our French master, 'Danny' Blakemore, was a vigorous proponent of the French interpretation of Franco–German history. He regularly distributed selected copies of dated French newspapers, ostensibly to improve our French but also to disseminate the strongly anti-German views that he held. As one of our texts for detailed study he gave us Daudet's *La Dernière Classe*, set in Alsace under the German occupation that followed the Franco–Prussian War and charged with patriotic emotion. The possibility of war once again began to be discussed, and it was from Blakemore that I first heard the phrase 'Cette fois, il faut en finir'. The Munich Agreement was concluded in the last term of my second year. The morning after the news was announced, our very conservative English master, 'Percy' Ingram, explained to us how peace had been saved and how laudable Mr Chamberlain's political stance had been. There was in the class a stocky flaxen-haired boy whose name, Hviesdoslav Mladek, we enjoyed massacring. His parents had come to Australia from Czechoslovakia and he had naturally received other views of Chamberlain's performance. When Ingram had finished his exposition, Mladek stood up at his desk and, white with rage, announced that while the Agreement might have saved England's skin, it was at the cost of Czechoslovakia, and that the whole negotiation was a piece of unmitigated turpitude. He was ordered out of the room but, as he marched defiantly to the door, he was followed by a growl of support from his class-mates. Czechoslovakia was soon replaced as a topic of conversation by Poland, about which nobody appeared to know anything. The newspapers were full of pretentious and, as we soon found out, totally uninformed assessments of the military strengths of the various European powers. There was general agreement that the British Navy was invincible and would prove decisive. For all but a few of us, Europe was very far away. War broke out as I prepared to take the Intermediate Certificate Examination.

A love of the Australian bush came very slowly. While I was at primary school my parents were much too busy scraping together a tolerable life to indulge in excursions into the countryside, and I cannot recall a holiday in the bush until I was twelve or thirteen.

There had been an earlier trip with my mother to Katoomba, a pretty resort in the nearby Blue Mountains, where she had gone for a few days to recuperate from some illness. I have a photograph of myself mounted uneasily on a horse, but my mother saw to it that, mounted or unmounted, I did not penetrate far into the bush. A little later another trip with my mother comes to mind, this time to Moree to which she had been attracted because of the reputation of the town's hot springs for alleviating rheumatism. Moree was then a typically sleepy Australian country town, and it was there that I first saw Australian aborigines in any number. But I do not recall that we even once left the town to admire the countryside.

Like the vast majority of Australians, I was a child of the city. However, a serious attempt to remedy this was made at one point, not by my parents but by a strange creature called Charles William Peck. How he came into our lives I am not quite sure. I think my mother met him somewhere, found him interesting, and invited him to dinner. The invitation was repeated and somehow drifted into becoming a regular fortnightly fixture. There was a Mrs Peck but we never saw her and he never spoke about her. Tall, silver-haired, and handsome in a conventional way, Mr Peck had the look and the style of a retired Guards officer. He had come out from England as a young man and had stayed in Australia, but we were never given a clear idea of how he had spent his life or what he had done for a living. Mr Peck brought with him the vision of a quite different world. At some stage in his life he appeared to have worked in the outback and knew a great deal about the flora and fauna of the bush. He also claimed to have lived among the aborigines and had written a book called *Australian Legends* which purported to be a collection of genuine myths recounted to the author by the aborigines themselves. The tales were of course apocryphal, but they had a certain charm and they were steeped in the atmosphere of the Australian bush.

When Mr Peck learned that my brother and I had not yet had a country holiday and knew virtually nothing at first hand about the real bush, he took it upon himself to remedy this. In those days unspoiled bushland was easily accessible from most parts of Sydney, and Mr Peck would take us out on day trips to interesting and beautiful places which he would further embellish with his own brand of slightly mysterious story-telling. Our first trip was to Botany Bay where, in addition to the bush, we were given a riveting account of the discovery and settlement of Australia. Mr Peck's visits stopped as suddenly and as incomprehensibly as they had begun, and I do not know what finally became of him. He told us once that he had left instructions that when he died he was to be cremated and his ashes were to be scattered around a white waratah. The waratah is a scarlet flower, and I have never met anyone who has seen a white one. As the family became more affluent, and my brother and I more independent, the foundations laid by Mr Peck were built on by long holidays in the Burragorang Valley and on the banks of the Hawkesbury River. By the time I left high school I had learnt to ride and felt altogether at home in the bush; but it was not until I had lived in England for some years that I realized that the smell of gum trees and the plaintive call of the Australian magpie had marked me for ever.

I entered the senior school in a blaze of General Proficiency, having obtained A grades in all subjects in the Intermediate Certificate Examination, including Theory of Music which I took as an extra. The two years spent in the senior school were a period of intensive preparation for the Leaving Certificate Examination which, for those who aspired to it, opened the way to a university education. When I entered the High, one was permitted to take up to ten papers at the Leaving Certificate Examination. The usual thing for those who aimed at the university was to take six or seven different subjects at ordinary level and three or four of these at honours level. Since, in the competition for university exhibitions, which remitted all fees, honours papers counted for more than pass papers, most candidates took six pass and four honours papers. By the time I began to prepare for the Leaving Certificate Examination the process of reducing the breadth of secondary education had begun and the number of papers that one could offer was reduced to eight. We were therefore limited to five or six ordinary papers and two or three honours papers. This meant that in the fourth year some restrictive decisions had to be taken, although they were not at once irrevocable. In general one chose to specialize on the maths and science side or on the languages and literature side. With an eye to the combination of degrees in Arts and Law then offered by Sydney University, I chose the languages and literature side, but I did not thereby abandon maths and science. Indeed, among the books that I won as prizes at the end of the fourth year there is one, a Modern Library Giant containing both *The Origin of Species* and *The Descent of Man*, which reminds me that I must have come top of the examination in physics.

My memories of the senior school are dominated by the figure of Hugh Brayden, the classics master. Because of his deep voice, slow meticulous delivery, and rather formal manner, he was known as the Morgue, but few of the boys had much contact with him. He did not teach in the junior school, and the number of boys who chose to pursue the classics to an advanced level was very small. In my year only two of us took Latin to honours level and no one took Greek. This meant that, especially in the final year, the advanced Latin classes became personal tutorials, and I was exposed to the influence of Hugh Brayden in a more direct way than is normally possible between teacher and pupil at school. In the junior school modern languages were taught in much the same way as Latin, the two essential principles being the acquisition of vocabulary and the avoidance of grammatical error. Of the two, avoidance of grammatical error was much the more important, and a faulty accent or a mistaken gender incurred a much greater penalty than a total insensitivity to the literary quality of a piece of verse or prose. Although in the senior school we did read much more French and German as literature, the virtues of the grammarian remained paramount. Curiously enough, it was the study of Latin that changed my perspective.

Hugh Brayden was a man of very broad culture. In addition to being an excellent classicist he had a profound knowledge of English literature and a good working knowledge of French and Italian. His Latin lessons rapidly ceased to be the exercises in literal translation to which we were accustomed and became instead sensitive explorations into the use of words. The precise significance of a Latin or English phrase was constantly illuminated by reference to some other language, and references to other languages became forays into other literatures. What we turned out for Hugh Brayden was judged by standards altogether different from those that applied elsewhere in the school: a grammatical error was simply a slip, and it was the literary quality of what we wrote that counted. As the Leaving Certificate Examination drew closer, the two candidates for Latin honours were offered additional tutorial sessions, and each Sunday morning I made my way to Hugh Brayden's home, as I had done five years before to Turnip Townsend's. He lived at Point Piper in an elegant batchelor's flat that looked out over the harbour, and it seemed to me then that I had never before found myself in so refined an environment. The walls were lined with books and interesting paintings, among which I was just a little shocked to see two nudes, and beside the gramophone, which was putting out a Mozart piano concerto as I entered the flat for the first time, there was a huge pile of records. Our Sunday morning lessons were a more intensive version of what we had at school, but there was usually time, either before or after, for a little open-ended conversation during which I managed to glean something of the background of the man. He had been born in Ireland and had received there a thorough classical education which was to have led him into an academic career. But trouble with his lungs had induced him to emigrate to Australia where he had settled for a quiet life as a schoolmaster, a job for which he was grossly overqualified. He appears to have spent his life pursuing his literary and artistic interests, moving gently within a small circle of friends who held him in the highest esteem. He was an immensely tolerant man, indifferent to religious or political argument, repelled by fanaticism of any kind. I owe him not only an inspired introduction to the study of language, but also an unforgettable glimpse into the life of a completely civilized man.

None of the other teachers made much of a mark. The discipline of the Latin classes spilled over into my work in French and German, but the English teaching was beyond redemption. It was assumed, quite wrongly of course, that we could all write English, and we were encouraged to produce florid and pretentious essays in literary criticism in which panache was given more credit than insight. I formed the conclusion while I was still at school that no specialist training was required to enable one to read the modern literature of one's own language, and I doubted whether reading this literature and talking about it constituted a serious academic discipline. I was highly amused when, many years later in Oxford, I heard exactly these views from the mouth of John Sparrow, Warden of All Souls. My bookshelves provide evidence that I must have been a pretty good linguist by schoolboy standards, for it appears that in my final year at the High I took the Dr F. W. Doak Prize for Latin, the Earle Page Prize for Modern Languages (*aeq.*), and the A. B. Piddington Prize for English Literature (*aeq.*) This I mention because the recitation of the names gives me pleasure.

My time in the senior school was a period of deepening gloom in the outside world. The British army had been swept out of Europe, and France had collapsed. Mr Chamberlain, of whom we had heard such great things, was replaced by Mr Churchill, and the school's foremost orator listened with admiration, but also with some scepticism, to his great speeches. It seemed altogether possible that Britain would go under and that Australia would then be left to its own hopelessly inadequate devices. We still talked in First World War terms of the British Navy blockading Hitler into submission, but we knew that we were whistling in the dark. The War was brought vividly home to us one day when Roden Cutler, an old boy who had lost a leg in battle and won Australia's first Victoria Cross of the Second World War, visited the school. We assembled in the Great Hall to do him homage and he made a stirring speech. But despite our overt confidence in Britain, we were all surreptitiously wondering whether the United States would take an interest in Australia's defence if Britain were defeated. This was a question that was fast becoming urgent, for Japan was adopting an increasingly aggressive stance, and even before the War there were many in Australia who regarded Japan as the real enemy. In my last month at the High, Japan attacked Pearl Harbour.

2

SIDERE MENS EADEM MUTATO

When I came up in 1942 the University of Sydney was a provincial backwater. Its motto *sidere mens eadem mutato* was I suppose intended to mean 'the same spirit under a different star', but it was never clear to me just what that spirit was presumed to be. The founding fathers were constantly engaged in internecine warfare, and the issue might well have been whether the spirit in question was to be that of Oxford or Edinburgh. None the less, for me the University opened vistas exciting beyond belief. To begin with there was the sudden freedom to do as I wanted with most of the day. I had enrolled in the Faculty of Arts to read, primarily, modern languages. This was not a decision, but the postponement of a decision. I do not think that I ever seriously considered the possibility that I would make my living out of modern languages. (Only a tiny handful of undergraduates in Australia then regarded their university education as anything other than a stepping stone to a career.) The prospect of later making the traditional move from Arts to Law remained open, but it had already lost some of its initial appeal. I simply did not yet know what I wanted to do with myself. It seemed a good idea in any case to spend a little time enlarging my general education while I made up my mind. This idea also appealed to my father who, like many intelligent men deprived by circumstances of a formal education, had a great respect for learning. The fact that all this learning was to be paid for by the public exhibition I had won made this high-minded choice essentially painless.

So I signed on for French 1, German 1, Psychology 1, and Elementary Italian. Italian was not taught in the secondary schools, and a crash-course was provided to bring you up to university level within a year, if you could stand the pace. That was Elementary Italian. I must explain the psychology. You were permitted to take four subjects in the first year. Having decided on the three languages, I had to choose between English, history, philosophy, and psychology for the fourth place. I took the view that history and philosophy, like English, were subjects I could read in my own time without any special technical training, and I hoped that psychology as set out in the syllabus might offer a harder centre. I was soon profoundly disillusioned, for psychology, as it was taught then and there, was a pot-pourri of twaddle; and although I was required to take the examination I stopped attending the lectures after the first few weeks. Instead, I sat in on all sorts of other lectures, especially those on logic given by John Anderson about whom I shall have more to say later.

Most of the lectures in Arts were given in the morning and the rest of the day was meant to be spent reading, preparing essays, or, for the linguists, cobbling together translations with the aid of dictionaries and grammar books. If you didn't take these scholastic exercises too seriously, you could knock them off pretty quickly, and that left a great deal of time for exploring other things that the university had to offer, of which the most important were human relationships. Very few of the Arts undergraduates were men. Students of medicine, engineering, and one or two other practical disciplines were exempted from military service because these subjects were judged to be of direct use to the war effort; but Arts men were called up at the age of eighteen, which most of them reached during their first year. Official versions of the history of the First World War tell us of the wave of warlike enthusiasm that moved young men irresistibly, but voluntarily, into the armed forces in 1914. The undergraduate entry into the University of Sydney during the years of the Second World War hardly suggests that there was a similar wave of enthusiasm then, for the faculties that conferred exemption were oversubscribed whereas the first year in Arts contained a mere sprinkling of men, hardly noticeable in the crowds of eager young women. In the upper years the number of men was reduced to perhaps twenty in all. Apart from those exempted on medical grounds there were about a dozen who were exempted on grounds of academic distinction. Each non-exempt faculty was allocated a small quota of reserved places for men who had done particularly well in the examinations at the end of the first year. These high-flyers were then permitted to continue their studies to graduation provided that their academic performance did not falter. What political or military considerations prompted this bureaucratic dispensation remain unclear to me. The young men who were thus privileged to study subjects that were not officially deemed to be of any importance to the War effort were naturally a rather unusual lot, made perhaps a little more so by the circumstances in which they found themselves. They formed closer associations with each other than they might otherwise have done, and it would not be an exaggeration to say that a certain *esprit de corps* prevailed. A string of names that I could recite would be familiar to any student of the intellectual life of Australia in the post-War years.

I thus found myself engaged by special dispensation in the pursuit of a liberal education at a time when the life of the city was dominated and changed irreversibly by the progress of the War. In the three months that elapsed between my leaving the High and entering the University, Japanese forces had overrun Hong Kong, the Phillipines, Malaya, Singapore, most of Burma, and the Netherlands East Indies, and they had begun the invasion of New Guinea. The collapse of Singapore and the casual elimination by the Japanese air force of the *Prince of Wales* and the *Repulse* sent out to defend it were events that changed Australia's history. The traditional mainstay of Australia's defence policy, in so far as it had one, was the assumption that Australia would be defended by the British Navy. We had learnt and accepted this doctrine at school, and so apparently had the politicians who were guiding the country's destiny. In the course of a few days it had become clear to everyone that the British Navy was powerless to defend Australia and that, without massive American help, it was indeed defenceless. Mr Curtin, the Prime Minister, in a historic speech, declared that Australia now looked to the United States and not to Great Britain, and thus initiated a process of psychological, political, and economic realignment that continues to the present day. Sydney rapidly filled up with American troops and within a few months the leisurely, conformist, provincial city of my childhood had vanished forever.

The academic life of the University, however, did not change so quickly. To my intense disappointment none of the modern languages I had chosen was taught in the humanistic tradition that Hugh Brayden had given me a taste of at school. Grammatical errors remained the most heinous of crimes, and the road to success lay in their scrupulous avoidance. We were set a small number of rather dull texts for detailed study, chosen not for their literary or historic interest but for their imagined usefulness as models of prose style. Verse was studied mainly as an exercise in the decipherment of linguistic complexity. Great store was set on the translation of the foreign language into English, and even greater store on the translation of English into the foreign language. Once a week we had a conversation exercise with a *lecteur* or *lectrice* for whom the language was native. One had the impression that all the courses were designed for aspiring school-teachers. At a later stage we studied some medieval texts — the *Nibelungenlied* and the *Chansons de Geste* — in the same dreary way. French was a good deal worse than German or Italian. The Professor of French at that time was G. G. Nicholson who enjoyed a formidable local reputation. His main interest was etymology and in his book, *Recherches philologiques romanes*, he endeavoured to provide strictly Latin derivations for a number of obscure words generally supposed to be of Germanic origin. We had to commit all these Latin derivations to memory, and it was only after I came to Oxford and had occasion to mention Nicholson's etymological work to scholars there that I learned that he was something of a laughing-stock.

We saw little of the professor of German, E. G. Waterhouse. He gave the impression of being an urbane and kindly man but, by the time I came into contact with him, he seemed to be more interested in the cultivation of camellias than in the study of German. By far the most interesting figure in modern languages was Victor Stadler, originally a lawyer who had fled Hitler's Germany and who was then employed by the University as an assistant lecturer in German. Stadler lectured to us on German literature which was for him a consuming passion. I got to know him fairly well, at least to the extent of occasionally visiting his home, and I was greatly influenced by his views and by his demonstration of what the study of literature could be. It was he who introduced me to the work of one of the few significant Australian poets, C. J. Brennan. Stadler regarded Brennan very highly and was attempting to translate some of his work into German. Although the tragic end of Brennan's life was acted out in that same quadrangle less than a generation previously, and although both Nicholson and Waterhouse had known him well, I never heard his name mentioned in lectures by anyone but Stadler. A couple of years later a friend of mine, whom I in turn had initiated into Brennan, undertook the laborious task of typing out all of Brennan's poetry from the corpus in the Sydney Public Library, for no complete collection of his work had yet been published. I received, and still treasure, a copy of that typescript.

My reaction to the stodge of language teaching to which I was exposed was ill-concealed rebellion. I read voraciously in all three of my languages but not at all along the lines advocated by my teachers (with the honourable exception of Stadler). I treated the compulsory linguistic exercises with contempt, but my continued presence in the faculty testifies to the fact that I must have been pretty good at them. Despite the system, I seem in the end to have acquired a good working knowledge of French and German and enough Italian to enable me to read the literature without effort. These were assets that provided me with a rich source of pleasure throughout my life, and I suppose I should be grateful to those whose unenviable job it was to teach rebels like me. I find it difficult; my principal emotion is still anger. But if the formal study of modern languages was a great disappointment, the exploration of other aspects of university life was not.

The dominant intellectual force in the University at that time was the philosopher John Anderson, about whom much has been written by his former students and associates. I find it difficult to assess his importance in a broader context, but in the constricted intellectual scene that Australia offered in his time he was a major revolutionary figure, and his influence was profound and lasting. To begin with, he was an overt, and perhaps deliberately provocative, atheist. (I recall his reaction to a notice posted by the Newman Society advertising a lecture on Eleven Formal Proofs of the Existence of God: 'If any one of them were any good, the other ten would be unnecessary!') Although overt atheism would not then have provoked even the raising of an eyebrow in most European universities, it was a perilous position for a university professor to espouse openly in Sydney before the War. Indeed, moves had at one time been made at the highest level to have him dismissed from his post, on grounds not very different from those used to condemn Socrates. Fortunately more rational counsels prevailed.

His philosophical position was a coherent system of uncompromising realism, a radical extension of the materialist metaphysics of his teacher, Samuel Alexander; but it also incorporated a curious amalgam of influences derived from Marx, Freud, and Joyce. In politics, his attitudes were vigorously anti-authoritarian. He argued that the excesses of those in power could be held in check only by unrelenting opposition to power on principle, by the public exposure of clandestine bureaucratic manoeuvres, and by constant warfare against all forms of obscurantism such as the censorship of books or other restrictions on free enquiry. These were real issues in the Sydney of those days where you could not freely purchase

The Decameron, Ulysses, or even some of the works of Freud. Anderson thus became the magnetic centre of what we would now call a dissident movement, and he was detested equally by the extreme Right and the extreme Left of the political spectrum.

The group that gathered about him in the University were known as Andersonians and their public forum was the Freethought Society which he had founded. Although by the time I entered the University I was already a committed realist, not too far removed in many of my attitudes from Anderson, and although I found his social stance attractive, I did not become an Andersonian. There were two reasons for this. The first was that I was not reading philosophy (at least not officially) and hence was not taken in as a member of the family; the second was that I found the group style of the Andersonians slightly ludicrous. Whereas Anderson was uncompromising in his opposition to the authority of other people, he did not welcome opposition, or even deviation, among his own disciples. Not only did the Andersonians adopt a uniformly Andersonian position on all major issues but, consciously or unconsciously, they adopted the characteristic style of his delivery and even some of his physical mannerisms. This was too much for me. But I must say that, of all the groups I encountered at the University, I felt most naturally at ease among the Andersonians, enjoyed their company, and formed lasting friendships with some of them.

It was through my friendship with a marvellous eccentric, Maxime Le Petit, that I came to know A. D. Trendall, the Professor of Ancient Greek. Max's French ancestry must have been very remote but, with a very swarthy complexion and large black sideburns, he managed to achieve a style that would not have looked out of place in Montmartre at the turn of the century. He was the best undergraduate classicist in the university and beneath the deceptive pose he harboured a deep and unaffected interest in the life and art of the Greeks. How he came to have this interest was a puzzle to me, for he was the product of an undistinguished State secondary school where a passionate interest in rather inaccessible art forms was unlikely to be readily generated or fostered. Whatever its origin, this interest was vigorously nourished by Trendall, who was the first example I had ever met of a man completely consumed by his subject. Trendall was already then immersed in the study of Greek vases in Italy, a subject on which he eventually became the world's foremost authority. He did his best, in an inclement environment, to generate more widespread interest in his subject: his lectures and the exhibitions he organized were an unexpected pleasure for the few who had the time and the inclination to stop, look, and listen.

Trendall had a good command of Italian, and it was he who introduced me to Silone. Although he was anything but left-wing, he mentioned in conversation one day, a little facetiously, that *Fontamara* was his favourite modern novel. I at once got hold of the book and consumed it in one draught; the opening lines were reward enough. Under the influence of Max and Trendall I began to teach myself Greek. Max lent me his copy of *Deigma*, the introductory text he had used at school. But I made little headway, being hopelessly over-extended by an ever-widening range of other interests. On the eve of a very promising academic career, Max suddenly developed jaundice and died within a few days of liver failure. As soon as I heard of his illness I rushed to visit him in his ward at the University hospital, the Royal Prince Alfred, but he was already comatose when I arrived. A scholarship was set up to commemorate his name, but I don't know whether it has survived the passage of time. I still have his copy of *Deigma*. I think he would not have minded my keeping it.

Of the various exotic growths that struck root in the University during the war years perhaps the most un-Australian was a group that eventually came to be known as the 'Poseurs' Push'. This was a collection of about a dozen young men from various disciplines who met once a fortnight or thereabouts in the home of one of them, usually on a Friday evening. A prepared paper would be read and then discussed; light refreshments brought the evening to an end. Gatherings of this kind were of course commonplace in the intellectual life of Europe, but I am not aware that they were ever a feature of Australian academic life. Who invented the name 'Poseurs' Push' is not known; it was certainly not one of the 'poseurs' themselves. It was, however, a deadly accurate reflection of what most other people thought of the group. Actually the 'poseurs' were very serious in their intentions and in their practice. It was at one of their meetings that I heard for the first time of Wittgenstein and the Vienna Circle. Logical positivism was not taught, or indeed even recognized as a serious subject, by John Anderson; and when attempts were later made by one or two heroic members of the 'Push' to discuss the subject with Anderson, his response was negative and fierce. On another evening I recall a vivid introduction to Toynbee and the problem of historicism; on yet another, a sceptical demolition of psychoanalytic theory. It may well be that the ideas discussed at those meetings would hardly have been avant-garde in London or New York; but each paper represented a personal exploration by a highly intelligent young man, and I can think of no part of my prolonged undergraduate experience that gave me greater pleasure than the meetings of the Push. As the original founding members drifted off, the Push came to an end; but in so far as I have been able to keep track of them almost every one of those young men did something interesting with his life.

With only a small part of the day committed to formal lectures, which in any case I attended less and less as time went by, I began to explore and to fall a little in love with the city itself. Sydney at that time contained a number of real bookshops, by which I mean those that sold something other than currently popular items and set texts. (When I last looked, in 1982, I couldn't find a single one.) In Campbell Street, then a genuine Chinatown where you could get a table d'hôte meal for next to nothing, there was the Advance Bookshop, run by an old anarcho-syndicalist who still carried, but I think had great difficulty in selling, dated political tracts of various kinds. I expect he made his living by providing his customers with books like *Ulysses* or *The Decameron* or *The Golden Lotus*. Banned books could readily be obtained in Sydney but they were naturally rather expensive. Had it not been for the subventions provided by University scholarships and prizes I might not have read them until some years later.

The Roycroft in Rowe Street, nestling at the side door of the then famous Hotel Australia, adopted a fashionable, slightly Edwardian air, but it carried a remarkable stock of serious literature, often in elegant editions. (I have a splendid India-paper edition of Wordsworth as evidence of almost unbelievable extravagance.) In Martin Place a rogue growth called the E.F.G. Bookshop had struck root, specializing in French and German books, but carrying in addition a curious assortment of scholarly items that you wouldn't find anywhere else. And there were several good secondhand bookshops where you could always find something of interest and once in a while a real bargain. I built up a collection of standard German classics from one such bookshop for a derisory outlay. During the unsettled conditions of the early war years, immigrants to Australia were often separated from their chattels. If these were not claimed in customs within some statutory number of days they were auctioned off. My secondhand book-dealer had taken to buying job lots of German books at these customs auctions and then offered the volumes for sale to his not very interested customers at a shilling apiece or less. For a shilling a volume I bought Cotta's 1858 edition of *Goethe's Collected Works*; for sixpence a volume a complete edition of Schiller. I subsequently learned that the *Goethe* had actually belonged to Victor Stadler, but he refused to take it back.

I spent a great deal of time wandering about the city from one bookshop to another, exploring interesting back streets; and occasionally, especially on those rare days when the city shone like a jewel in the clean air, I might pass the whole of an afternoon in some deserted corner of the Botanical Gardens, doing nothing more constructive than watching the small craft scud across the harbour. But it was at night that the most durable fetters were forged. If you have never walked out into the night air in shirt sleeves and been assailed by the combined efforts of oleander and frangipani blossoms, you won't appreciate how powerful nostalgia for a warm-weather country can be. In Sydney in November the nights can already be hot enough to drive you out of doors; often I would get up from my books at the close of the evening and walk for miles through sleeping streets, their fragrance sharpened by the night air, and overhead a sky so crowded with bright stars that to a European eye it would have looked artificial. If I took the road over the hills that sloped down to the harbour, I would have glimpses of the indigo water flashing with reflected shore lights; and if I chose a road that followed the sea front I would hear the surf running as I neared the beaches; and if a sea breeze sprang up there would be the smell of salt water on the air. Forty years ago you could walk alone anywhere in that city, see hardly a motor car, meet no one but an occasional wanderer like yourself, and come home with the dawn reeling in a turbulence of sights and sounds and smells and images that have remained undimmed by distance or by time (Plate 2).

During my first eighteen months at the University the fortunes of war had undergone a dramatic change. The German armies had been destroyed at Stalingrad and were being remorselessly pushed back elsewhere on the Russian front; the American fleets, in the naval battles of the Coral Sea and Midway, had removed the possibility that the Japanese might invade Australia. It was clear that, with the massive growth of American military power, the Allies were going

Plate 2. Double Bay, Sydney harbour, in the days of my youth. The hills that sloped gently down to the water's edge were studded with old houses set in large gardens. The harbour front has now been sacked and encased in a wall of high-rise apartments.

to win the War in the end, although no one was under any illusion about how long it might take to achieve that end. It was thus possible to think, at least in the long term, about a post-War future. I had made up my mind by the middle of my second year that I had had enough of Arts, and it was clear that I would have to make some decision pretty soon about what I proposed to do next. There was a special arrangement in the University at that time, and perhaps there still is, that permitted you to acquire a formal B.A. within two years, instead of three, provided you had accumulated enough courses and then proceeded to a second faculty. What induced me to transfer to medicine rather than law remains obscure to me even now.

It is easier to define factors that were not material considerations than to give an unromantic account of what my motives actually were. I was certainly not moved by the financial inducements of medical practice. Although in Sydney, then as now, the practice of medicine was a pretty sure road to affluence and social standing, these were simply not components of the set of values that moved me at that time. Indeed, quite the reverse. Nor was I driven by humanitarian motives such as the desire to heal the sick or otherwise alleviate the misery of the human condition. I think I took a pretty cynical view of those who professed motives of that kind. My interest in science did not disappear completely during the period of my linguistic studies. I had another go at *The Origin of Species* and this time managed to read it right through. I remember saying at one of the meetings of the Poseurs' Push that I thought natural selection might well be the greatest idea any man ever had. I still think so. I read and enjoyed what were actually bad biographies of Pasteur and Marie Curie. I had looked hard at what was on offer in the wartime faculty of law and rather feared that it might be even more pedestrian than what I had received in Arts. What medicine had to offer was by no means clear, but I felt that at the very least I would be exposed to something new and in principle interesting. But I think my main (and scarcely credible) reason for choosing medicine was a vision that I had of myself as some kind of future Chekhov or Schnitzler or Döblin.

Immersed in the study of literature, I had begun to harbour ambitions as a writer and, apart from some passionate love lyrics, had actually produced a few pieces that were perhaps not absolutely hopeless. A short story and an article found their way into print in *Arna*, a rather pretentious literary magazine produced by the Arts undergraduates. Chekhov was an overwhelming experience, and it was his short stories that first aroused in me an interest in the doctor as a writer. I had read a good deal of Schnitzler, with greater enthusiasm then than I have for his work now; and *Berlin Alexanderplatz*, which I happened to find in a secondhand bookshop, introduced me to Döblin. I thought Brown's *Rab and his Friends*, which I first met in an anthology, a masterpiece, sentimental though it was. In all these writers I was fascinated by the way the medical experience was distilled into a vision of reality that had a dimension not easily achieved otherwise. I began to dream about accomplishing something like that; and the idea of reading medicine, primarily to inform my writing, began to take hold. Improbable though it may seem, there is no doubt that I regarded this implausible vision of the future as a perfectly practical proposition: my activities during the first two years of the medical course provide tangible evidence that I was taking my dream seriously. To my medical studies I devoted only the perfunctory attention that was necessary to ensure a pass in the annual examinations, but I produced the beginnings of a novel, a long treatise on aesthetics and various smaller items that my filing cabinet still retains in manuscript form.

When I applied to enter medicine my attention was drawn to the existence of a marvellously far-sighted benefaction, the Struth Exhibition. This provided fifty pounds a year for a graduate in Arts proceeding to the study of medicine. I was interviewed by the Vice-Chancellor, Robert Wallace, who said some very kind things to me, and I left his study in a state of high elation and unprecedented affluence. I do not to this day know who Mr Struth was; but for six years his foresight provided me with a steady stream of banned and other books.

Before I begin to describe my time as a medical student I must take a deep breath. If I had been disappointed in what I had been offered in Arts, let me say that it was manna from heaven compared with the nightmare of the first two years of the medical course. To begin with there were the numbers. As I have described, the wartime Arts course was severely depleted, but medicine was massively oversubscribed. No quota had yet been established to limit the entry into medicine, and I believe that at one point there were something like eight hundred students in the first year. It was tacitly understood that the rate of failure and hence of elimination from the course would have to be such as to reduce the number to about two hundred by the end of the third year, for this was the maximum number that the teaching hospitals could accommodate. All classes were duplicated or triplicated but were hopelessly overcrowded none the less, and you could not help feeling a little like Charlie Chaplin in *Modern Times*, pinned to the production line. But it was not so much the overloading of the teaching facilities that sapped the course of any intellectual content; it was the decadence of what was taught.

In the first year we had to contend with physics, chemistry, botany, zoology, and the beginnings of anatomy; in the second with the full force of anatomy and an introduction to physiology and biochemistry. The physics and chemistry were little more than a recapitulation of what had been taught in the upper years of the school curriculum, with a small dash of organic chemistry at the end. Botany and zoology consisted almost entirely of detailed descriptions of the morphology and life-cycles of various plant and animal forms, accompanied by practical classes in which our main task was to produce elaborate drawings of specimens which we had either mauled ourselves or which had been prepared for us by more competent hands. I did not at any time hear Darwin or Mendel mentioned, and if their ideas had indeed percolated into those dreary departments they must have been reserved for more advanced classes to which medical students had no access.

Anatomy was the ultimate barbarism. Essentially everything in the medical course was taught and examined as an exercise in rote learning, but it was in the anatomy department that feats of memory reached their apogee. The anatomy teachers saw themselves as lineal descendants of the great Edinburgh anatomists and their main aim in life seemed to be to ensure that no detail of the anatomy of the human body remained unmemorized. There were not enough cadavers to go round, so that the dissection was done in layers: one group of students would take off an inch or so and would then be replaced by another group who went an inch deeper, and so on. Since no one dissected a cadaver completely, salvation lay in committing Cunningham's *Textbook of Anatomy* to memory, which, astonishing as it may seem, actually was achieved by a few of the most diligent students. It was a department in which enormous prestige could be gained by identifying in flight the small bones of the wrist, tossed high in the air.

Our introduction to physiology was remarkable even for the Sydney Medical School. There was a set text (Winton and Bayliss) which we were required to purchase and bring with us to each lecture. The lecturer would go through each chapter page by page, laboriously dictating additions and emendations. If you wished to do well in the examination you had to commit the purified text to memory and reproduce the relevant bits on the day. I have little recollection of what we were given as Introductory Biochemistry, perhaps because I didn't often go to the lectures, but, as far as I can recall, the course dealt mainly with nutrition and with vitamins in which the professor was said to have an interest.

By the time I reached the third year I was not at all certain that I would have the patience to complete the course. At each examination I had done the minimum necessary to ensure that I was not eliminated, but it was only my non-medical activities that interested me and I had almost decided that in choosing medicine I had made a mistake. My mind was changed by an unusual book. The study of the nervous system was reserved for the third year, and I had at one time thought that the nervous system was something I could get interested in. I no longer had any such expectation, for the teaching in the third year was done more or less by the same people and certainly in the same way as in previous years, and I found the prospect of yet another year of this infinitely depressing. I cannot now remember exactly how John Fulton's *Physiology of the Nervous System* came into my hands. It was not a set text, and indeed professional neuro-physiologists regard it as a rather wayward book. I began to thumb my way through it in a very detached way but, as the pages flew by, my interest and enthusiasm increased. Unreliable it might well be, but it was unlike any of the texts that had previously come my way. The information was presented in a tentative manner and against a background of experimental evidence. The evidence was discussed, its strength expounded, its weakness exposed. The book had a sense of history as well as a vision of the future. When I had finished it I found to my astonishment that I had caught fire.

I began reading more widely in physiology, mainly monographs that had caught my attention or that had been referred to in Fulton's book. It was dabbling rather than systematic reading, driven entirely by my own curiosity. I also took in some bits of biochemistry, the selection again being made in a completely haphazard way by the changing focus of my interest. I began to devise experiments (thought experiments of course) to resolve all the great outstanding problems that I came across in my reading. I must somehow have succeeded in the midst of all this ferment in acquiring a good deal of information that was relevant to the formal curriculum, for in the examinations at the end of the third year I obtained very high marks in physiology and biochemistry, in physiology the highest marks possible. This result was greeted with incredulity by my undergraduate contemporaries, for the leading lights in the year, the great memorizers, had already established themselves by this time and I was certainly not among them.

At the end of their third year medical students were transferred from the Old Medical School, a slightly crazy Victorian Gothic building situated close by the main university quadrangle, to the New Medical School, a sedate rectangle of red brick erected by the generosity of the Rockefeller Foundation and joined by a causeway to the Royal Prince Alfred Hospital. This move marked the transition from pre-clinical to clinical medicine. I now viewed the prospect of clinical medicine without enthusiasm. I rather feared that I would be dragged away from the genuine interest in physiology that I had developed and would be subjected once again to the mindless treadmill of competitive rote learning. I did my best to stave off the evil day.

During the long vacation between the third and fourth years I began what were intended to be my first scientific experiments. A few years previously a distinguished group of neurophysiologists, Jack Eccles, Bernard Katz, and Stephen Kuffler, had been active at the Kanematsu Institute of the Sydney Hospital. The War had dispersed them but some of their electrophysiological equipment had found its way into the New Medical School where, as far as I could see, it remained idle. The man who appeared to be in charge of it was at that time training to become a neuro-surgeon and had his hands full with other things. I asked him none the less whether he could find the time to show me how to use the equipment, which I hoped would enable me to test certain theories I had formed about the nature of visual after-images (images that persist after you've closed your eyes). He did indeed show me what knobs to turn and, with the help of one of the animal technicians, how to anaesthetize a cat. For several weeks I shone sequences of red and green lights into the eyes of cats and accumulated a great pile of electrical recordings taken from the backs of their heads over where I thought the visual centres of the brain ought to be. All these records were of course quite uninterpretable, but this does not seem to have deterred me.

One day A. K. McIntyre, a very competent neuro-physiologist who happened to be visiting the university, called on me. Someone had told him about my capers and I think that he was perhaps curious to see what was going on. When I had explained my experiments to him he asked me very gently why I had chosen to work on the cat. My explanation was simply that the cat seemed to be what most neuro-physiologists worked on. 'That might well be so' he then said, even more gently, 'but there's no use shining red and green lights into cats' eyes because cats are colour-blind'. The acquisition of that interesting piece of information brought my career as a neuro-physiologist to an end.

The fourth year was not quite as bad as I had feared. There were three subjects: pharmacology, pathology, and bacteriology, which included immunology. Pharmacology presented no problem for it was simply an extension of physiology with the addition of a few complex chemical formulae. But pathology was a caricature. It was the exact counterpart of the teaching in anatomy, except that here it was the morphology of diseased tissues, as seen in bottles in the pathology museum or in sections of tissue examined under the microscope, that had to be committed to memory. It remains a mystery to me how I succeeded in satisfying the examiners in this subject for I distinctly remember that in the practical examination I diagnosed a section of an inflamed appendix as a cancer of the stomach. On the other hand, the Professor of Bacteriology, H. K. Ward, I remember with affection and gratitude. He was I think the only medical professor in Sydney in my time who had done scientific work that is still remembered, and in a completely unforeseen way he was to play a decisive role in determining the course of my life. The great bacteriologists had been heroes of mine since school-days and I viewed the prospect of learning a little about the subject with cautious enthusiasm. The enthusiasm soon evaporated.

Ward's lectures, delivered in a dry unmodulated voice, were essentially a form of dictation in which each sentence was enunciated slowly twice so that we could have time to take it down. It was difficult to reconcile the bored old man who lectured to us with the tales that we had heard about his youth. He had been a Rhodes Scholar at Oxford, had rowed in the boat-race, and had apparently done excellent work at the Sir William Dunn School of Pathology. He had then gone to Harvard where he worked in the department of the celebrated bacteriologist, Hans Zinsser, rising rapidly to become an associate professor. It was at Harvard that he did the work that is still occasionally quoted. The story went that when he returned to Sydney as Professor of Bacteriology he had tried very hard to change the style of medical teaching, but had been frustrated by his colleagues. This may well have been true, for he certainly had the air of a disillusioned man. But not always. There were moments in his lectures when he broke off the tedious repetition of his sentences and spoke freely for a few minutes about experimental work that he obviously admired. And he always seemed very pleased to discuss experimental ideas with anyone who sought him out for this purpose. In the course of my usual haphazard reading around topics that happened to have caught my attention, I became interested in bacteriophages (viruses that infect bacteria); and when after one of his lectures I asked him whether they might not be used therapeutically, I was regaled with a marvellous half-hour of learned enthusiasm.

Each year Ward made one public gesture, without parallel in my experience of the Medical School, that showed his fundamental allegiance to experimental science. The examination paper in bacteriology was the only one that contained a question that could not be answered by the exercise of memory alone. There was an endowed prize in bacteriology, the H. E. Waldron Memorial Prize, and if you wanted to be considered for that you had to attempt an optional question that was always experimental in nature and required some ingenuity to answer. It still gives me satisfaction to recall that I won the H. E. Waldron Memorial Prize. Ward did not forget it either, for three years later he introduced me to Florey and was thus directly instrumental in my obtaining a scholarship to Oxford.

The last two years of the medical course were given over entirely to clinical subjects: medicine, surgery, obstetrics and gynaecology, psychiatry, and the minor clinical specialities. None of this interested me in the least except for some aspects of medicine, which I found fascinating to read about but not to practise. To psychiatry I reacted in much the same way as I had done some years previously to psychology: the subject as presented to us was verbose and empty. Most of the clinical teaching was done by the physicians and surgeons at the teaching hospitals, in my case the Royal Prince Alfred to which I had been posted. With a few memorable exceptions, these clinicians were a pretty unattractive lot—affluent, and hence complacent, contemptuous of theory or indeed of anything that smacked of scientific sophistication beyond their reach, pretentious not only in medical matters but also in all sorts of other subjects about which they usually knew nothing. I contrived to avoid as much of this posturing as the system would permit and spent most of my time attempting heroic experiments in a wonderfully spacious laboratory that had been placed at my disposal in the Department of Medicine. How this miracle came about is of more than personal interest for it sheds a vivid light on the extent of scientific activity in the Medical School in those days.

C. G. Lambie, the Professor of Medicine, was a very small, very learned Scot, who spent his years of exile in Sydney in a state of ill-disguised warfare with the powerful clinicians at the Royal Prince Alfred Hospital who disliked both the man and his methods. When his exile was over he worked his way back to Scotland as a ship's doctor and was not heard of again. He was a great systematist, organizing all medicine into neat categories composed of endless subdivisions. While I was a student his massive two-volume work on clinical diagnosis appeared, a great Gothic structure that was much too systematic to be of any value in the practice of medicine.

The clinical students in Sydney had to read it; I doubt if anyone else did. Lambie had no personal experience of experimental science, but unlike many of his clinical colleagues he was not against it. I approached him with my request for laboratory space essentially because more space was allocated to the Department of Medicine than to any other department, and as far as I could see most of it was unoccupied. What I wanted to investigate was whether the course of an auto-immune disease (a disease in which the patient makes destructive antibodies to his own tissues) might not be modified by the administration of antibodies directed against the patients' own antibodies. Recent developments in immunology make this seem, retrospectively, a rather advanced idea, but there was no real hope at that time that anything useful could be done along these lines. None the less experimental animal models for some human auto-immune diseases were available, and I desperately wanted to have a try.

Lambie warned me that this might seriously deflect me from my clinical studies, but he did not turn me down. On the contrary, he offered me the sole use of a laboratory that could comfortably have housed half a dozen workers. It was one of several such laboratories that were available. To help me with the work I managed to interest a friend who had already taken a degree in chemistry and who was then reading medicine a couple of years behind me. In one of the papers that had influenced my ideas on auto-immune diseases, ducks had been used to prepare an anti-serum. So I had to have ducks. Unfortunately none of the animal houses to which I had access were equipped to accommodate ducks, but I managed to wheedle out of the authorities permission to use a small outdoor pen which must at one time have been used for something but which was now totally idle. Into this we deposited four unprepossessing ducks, purchased cheaply.

A few days later we turned up with the concoction I proposed to immunize them with and set about the task. It proved not to be easy, for it was some time before we managed to catch one of the birds and hold it in a position that permitted injection into the peritoneal cavity, which was what I intended. When I finally succeeded in making the injection the bewildered duck gave a little cough and promptly died. I assumed that it must have died of exhaustion produced by its efforts to escape us; but when the second duck did exactly the same thing I decided it was time to visit the Veterinary

School and make some enquiries about the anatomy and physiology of ducks. There I learnt that birds have air sacs which are of interest in the present context because they communicate directly with the lungs. It was explained to me that in all probability what I took to be the peritoneal cavity was an air sac and that I had no doubt suffocated the birds by injecting my concoction into their lungs. I did better with the two remaining birds. When our experiments were completed (they were, needless to say, a complete fiasco from beginning to end), I presented Lambie with a written report on the work. Instead of throwing it into the waste-paper basket where it belonged, he actually encouraged me to publish it, with the result that in the middle of my final year as a medical student my first scientific paper appeared in the *Medical Journal of Australia*. I am heartily ashamed of it.

The last few months of my undergraduate career were a race against the clock. Somehow before the final examinations began I had to acquire, pretty much from scratch, at least the fundamentals of surgery, obstetrics, and gynaecology. Medicine, with a little luck in the examination paper, I could cope with. Those months flew by in a nightmare of unremitting memory work, and I look back on them as the least pleasant episode in an unpleasant medical education. Luck must have been very much on my side in the final examinations for I managed to graduate with a good enough honours degree to be accepted as a resident medical officer at the Royal Prince Alfred Hospital. Apart from the scions of notable medical families already established there, this was a privilege normally reserved for only the most diligent memorizers.

My time as a resident medical officer at the Royal Prince Alfred Hospital was more fun than I expected. Although living at home had not placed great restrictions on my activities, living away from home, with a modest sum provided by the Hospital authorities to serve as pocket-money, conferred a new sense of personal freedom. It was more a sense of freedom than the real thing, for we were kept extremely busy, often working round the clock. I found that I didn't dislike treating patients as much as I thought I would. It was a relatively easy, intellectually unexacting task, and if you were reasonably conscientious you were rewarded by the feeling, probably exaggerated, that you were extremely useful. I am told by those who remember me then that I was a good resident who, if he hadn't been misguided, might have become a successful clinician. However, my desire to try my hand at medical research was by then firmly rooted and the prospect of moving up the conventional clinical ladder was without attraction.

My first assignment was to be houseman to the hospital's senior surgeon who had a reputation for great ferocity. His behaviour and his surgery were certainly very eccentric, and I later learned that he must already then have been suffering from the early symptoms of a brain tumour which led a couple of years later to his death. I have never had any problems in getting on with eccentrics and, to everyone's surprise, the surgeon and I hit it off very well. Indeed, we actually wrote up a case together for the *Medical Journal of Australia*. (This I do not include in the list of my publications.) There was a great deal of bragging in the Hospital about its pre-eminent role in world medicine, but since I did not see this reflected in the medical literature or in the comments of the overseas visitors who occasionally passed through, I formed the conclusion that the Hospital's view of itself was not necessarily shared by the rest of the world.

My next posting was with a fashionable physician from whom I learned nothing. While I was with him I received my first offer of a job. It came from R. D. Wright, universally known as Panzy Wright, Professor of Physiology in Melbourne. Panzy had come to hear about me through Alf Conlon, a celebrated figure who had been associated with the University of Sydney for years. During the War Alf had had a meteoric career rising from the position of University Manpower Officer to become the Director of Research and Civil Affairs for the whole of the Australian armed forces. I met him when he returned to the University after demobilization to complete his medical studies. He had the air of an elder statesman and gave the impression of being on first-name terms with virtually everybody of importance in Australian political life. In any case he did know Panzy Wright who at that time was on the look-out for young scientists who might contribute to the research in his own department. Largely due to the work of F. M. Burnet, Melbourne enjoyed a better reputation as a centre for medical research than Sydney; and in our conversations Panzy made much of the differences between the two places. I later found that, Burnet apart, these differences were not as great as they seemed. None the less there were certainly some interesting things going on in Panzy's department and, as there was in any case nothing else on offer, I accepted his job. We agreed that

I would go to Melbourne when I had completed my round as a resident medical officer.

In a corner of one of the basements of the Hospital there was a mysterious area known as the Clinical Research Ward. This was directed by a young man named Ruthven Blackburn, a son of Charles Blackburn, an eminent physician who finally became the Chancellor of the University. Ruthven Blackburn had recently returned from the United States and I was very curious to know what was going on in his domain. I asked to be posted there in the next shuffle of the housemen, a request that was granted without difficulty for the Clinical Research Ward was by no means a coveted posting. Blackburn's main interest appeared to be in certain blood disorders, but a variety of other puzzling cases also found their way into the ward. There were fewer beds there than in any of the other wards in the Hospital and the cases admitted were investigated in much greater detail. I picked up a little haematology and rather more clinical biochemistry, but also managed, with Blackburn's encouragement, to spend a good deal of time in a small annexe that served as a laboratory. There I tried without success to produce stable changes in the shape of red blood cells in the hope that this might throw some light on the physiological significance of their curious natural shape.

While I was busy with red cells a case was admitted that moved my interest vigorously in another direction. This was a man with total kidney failure who had produced no urine for about a week. In those days such patients died unless they recovered some kidney function, and the main aim of treatment was to keep them alive until this recovery occurred. The failure to produce urine caused very interesting changes in the blood chemistry and especially in the concentrations of the common salts in the blood. However, the only reliable way to measure these salts was with a new device called a flame photometer, and the only one of those in Australia at that time was in Panzy Wright's department in Melbourne. I rang Panzy about the problem and at his suggestion flew down to Melbourne with the samples of blood and made the necessary measurements. I did not manage to save this patient's life but I did succeed in keeping him alive without any kidney function for an unusually long time. A couple of days after he died I received a visit from his widow who had brought me a snake-skin wallet as a gift. I was naturally reluctant to accept it at first but was finally persuaded by an irresistible argument: 'Please take it, doctor. You don't know how grateful I am. For a while there, I thought he was going to survive'.

Because the management of this case had been so informative, Blackburn encouraged me to present it at Grand Rounds, a weekly gathering where unusual cases were presented to an assembly of the whole Hospital staff, or at least those who were interested and free to attend. I put some work into this presentation, and, for the first time in its history, the hospital was shown diagrams of the changes that had taken place during the course of the illness in the concentrations of the salts in the blood. (Such diagrams were well known in other parts of the world as Gamble diagrams, after J. L. Gamble, an eminent professor of paediatrics at Harvard who had introduced them years before.) My efforts met with a predictable response. As soon as the presentation was finished, a senior physician stood up and announced that while all this clever work had been going on in the Clinical Research Ward he had also been treating a patient with total kidney failure in his ward. He, however, had merely rendered the patient's urine alkaline with a little potassium citrate and, whereas the patient in the Clinical Research Ward had died, his patient had recovered. Urged on by Blackburn (to his eternal credit) I stood up and asked, as meekly as my inflamed state would allow, how this eminent physician had succeeded, with a little potassium citrate, in rendering alkaline a urine that did not exist. There was a long silence, and then the junior medical staff brought the house down. This incident is still remembered at the Royal Prince Alfred Hospital, but it would certainly have put paid to my chances of a career there if I had had any ambitions in that direction.

After the Clinical Research Ward I was allotted a mixture of anaesthetics and casualty duty. This involved a great deal of running around but very little cerebral activity, and I was rather enjoying it as a form of sport. One morning, when I was about to put a nice old lady to sleep prior to surgery, I received out of the blue a telephone call from H. K. Ward, the Professor of Bacteriology whose prize I had won three years previously. 'Would you like to meet Florey?' he asked. Those were the exact words, for they are graven in my memory. 'I have him here in my room and he'd quite like to talk to you'. Florey and the Oxford contribution to the production of penicillin were already a medical legend and for a moment I found it difficult to assimilate the idea that I was actually being offered the opportunity of talking to the great man. 'Do you mean *the* Florey?' I asked. 'Yes' said Ward, 'Florey'. I was on duty, but I managed to shake a little time free and tore across to the Bacteriology Department in the Medical School.

There was nothing dramatic about Florey's personal appearance, none of the aura that romantic biographers insist on imposing on eminent men. He was very conventionally dressed, his greying wavy hair neatly parted down the middle. He looked like a moderately successful business man, and the only thing about him that did not fit that image was a pair of plain, round, steel-rimmed spectacles that gave his eyes a rather piercing look. (I later learned that these were standard National Health Service issue.) He was at the time trawling for young Australians whom he could recruit to the John Curtin School of Medical Research in the newly established Australian National University at Canberra. He had a rather harsh, metallic voice and his speech was very direct, almost laconic: 'Professor Ward tells me you might be interested in doing experimental work'. I told him what I had done and what I hoped to do. He asked a few questions about what I had learned as an undergraduate and did not seem too impressed with what I told him. I think he must already have made up his mind about me from the information Ward had given him, for he came straight to the point: 'How would you like to come to Oxford?' It was like asking a starving man whether he'd like a meal. My slightly incredulous reply must have amused him, for a smile flickered across his until then impassive face. I went on to explain, however, that I had already accepted a job with Panzy Wright and that I could not therefore go to Oxford straight away. This seemed to pose no problem. Florey thought it might be a good idea for me to learn a little more about physiology for a year or so and then come to Oxford. He undertook to obtain a travelling scholarship for me. And so it was arranged. The interview had taken about ten minutes. I thanked him, said goodbye, and walked out into the bright sunlight.

With my future taken care of for the next few years, I could see no point in continuing to mark time at the Royal Prince Alfred Hospital and I gave notice that I would leave at the end of my current assignment. This decision caused some surprise, for young doctors did not normally leave that hospital on their own initiative; they left when the Hospital decided that their services were no longer required. I can't say that my time there was time wasted for I had, despite myself, learnt a little about clinical medicine and I had acquired, directly out of the clinical experience, some serious new interests. But the main residue of my hospital experience was the friendship of some remarkably loyal and attractive people. As a medical student I had not formed particularly close associations with any of my colleagues, but when that selected group of us took up residence in the Hospital and saw each other, day in, day out, in all sorts of circumstances from the hilarious to the harrowing, personal relationships began to be explored at a much deeper level. I made some friends then who have remained friends despite the great distance that eventually separated us and the rarity with which we came together. This handful of doctors, now all successful in their various ways, have watched my progress over the years with a benign eye and have never failed to celebrate with some moving gesture any success of mine that came to their attention. Their friendship has remained one of my strongest links with Australia, and I thank them for it.

When the other housemen heard that I was leaving the Hospital they decided to have a farewell party. This involved breaking a number of Hospital rules including the strict injunction that alcohol was not to be consumed on the premises. A keg of beer draped in a white sheet was smuggled on a Hospital trolley into the resident medical officers' quarters, and after dinner the great Australian party got under way. It was decreed that I had to down a glass of beer with each of them. I did my best, but half-way through the ritual I passed out and was carried to bed. Thus ended my clinical career.

During the seven years that I had spent becoming a Registered Medical Practitioner in the State of New South Wales momentous changes had taken place in the world. The War had been won, the atomic bombs had been dropped, and the liberation of the concentration camps had revealed entirely new dimensions of human cruelty. Australia's traditional political affiliations had been severed, and the swelling tide of immigrants was beginning to transform the texture of Australian social life, at least in the cities. Ham-and-beef shops were making way for delicatessens; restaurants of various national flavours were sprouting everywhere; and string quartets began to be heard even in the suburbs. But none of these things produced so profound an effect on my life as the advent of Alexandra Brodsky.

I first met her at a party at the home of one of my aunts. Domestic parties played a major role in the social life of young Sydney. At one end, they were remarkably well-behaved, middle-class, and, in tempo, middle-aged affairs, at which one ate a great deal, drank very little, and listened with half an ear to classical music. At the other, they were self-consciously bohemian riots at which young men attempted, almost always without success, to win the favours of unattractive young women against a background of too much beer and poor jazz. The parties organized by my aunt were at the extreme right-wing of the well-behaved variety. Proceedings were well under way, and I was making conversation in a corner, bored out of my mind, when the front door opened for a late arrival. A slender girl walked in wearing a green dress under a dark-brown coat, with blond hair piled on the top of her head. The evening was transformed. When it ended, I asked whether I could see her home. A few weeks later, I asked her to marry me. As I look back on our years together, I can only conclude that, despite my difficulties with the idea of God, some marriages are none the less made in heaven.

3

INTERLUDE IN A SOUTHERN CITY

We spent the first few days of our married life at Jervis Bay, a small resort about a hundred miles down the coast from Sydney. It was then little more than a long empty beach surrounded by uninhabited bush (Plate 3). The one hotel was called, with shameless exaggeration, the Naval Lodge because of a nearby naval station which was not actually in use at the time. When we returned to Sydney we at once made ready for our move to Melbourne where the job that Panzy Wright had arranged for me was waiting. We left by the night train, the Spirit of Progress, which would take us as far as Albury on the Victorian border. There, in the early hours of the morning, we would have to change trains, for the railway gauge in the State of Victoria was different from that in New South Wales. As the train rolled slowly out of Central Station and we settled down for the long journey, images of my childhood and of

Plate 3. Alexandra at Jervis Bay, December 1950.

my school and university days crowded into my head; but I do not think that I yet envisaged the possibility that I might be leaving for good.

In the cities of Australia in those days of acute housing shortages and unscrupulous rent rackets it was no simple task, unless you had the means, to find somewhere to live. In this we were greatly helped by the enterprise of some of the younger members of Panzy Wright's staff. What they had found for us was the front part of a modest bungalow in Balwyn, a clean new middle-class suburb a few miles from the centre of the city. The house was owned by a man called McConnan who was a district bank manager. His wife and daughter had gone off on an extended European tour, something that a bank manager's family could then afford to do perhaps once or twice in their lives, and McConnan had decided to let off the major part of the house while he and his son confined themselves to the two back rooms. There was only one snag: he wanted a month's rent in advance and I didn't have that much. I asked Panzy Wright whether the University of Melbourne might not let me have one month's salary in advance, but Panzy did not think this could be done. Instead he wrote me out a personal cheque for the amount required. It was the first of many kindnesses that I received at his hands. McConnan turned out to be a modest, decent, rather sentimental man, and we owe it to him that the beginnings of our married life were not the nightmare that usually confronted young couples starting out then in furnished digs. Each morning Alexandra and I caught the tram that ran along the Balwyn Road to the city, and each evening we made our separate ways back along the same route. At weekends we explored Melbourne and its surroundings. Comparing Melbourne with Sydney is a well-known Australian pastime. In terms of outward appearances, it was in those days rather like comparing Manchester with San Francisco, but Melbourne, though smaller, was very wealthy, and some of its citizens had a highly developed sense of civic pride. This showed itself in a number of well-endowed institutions like the National Gallery of Victoria which housed a remarkable collection of European as well as Australian paintings. Since the beaches were further away and we had no motor car, we spent less of our time soaking up sunshine than we would perhaps have done in Sydney, but it didn't strike us as a change for the worse.

The Physiology Department at the University was a lively place. Panzy ran it in an overtly paternal fashion, but the relationship between him and even his most junior staff was very informal. He had nothing of the remoteness that I had found in some of the Sydney medical professors. Stocky, powerful, with a shock of grizzled hair and massive black eyebrows, he affected the style and the speech of a rural simpleton, but he had a sharp eye for talent and he was devoted to the idea of making Melbourne a world centre of medical research. He made friends and enemies easily, but with time his enemies forgave him and his friends, a large class that included most of the staff of the Department of Physiology, remained loyal. Completely unconventional, often deliberately uncouth, he played a central and colourful part in the vigorous social life that surrounded the laboratory. He made no attempt to direct my research but encouraged me to pursue my new-found interest in kidney failure. This brought me into close association with Derek Denton and Ian McDonald, the principal members of the loosely knit team whose research centered on problems of this kind. Denton was already then an impressive entrepreneur, but while I was there most of the bench work was being done by McDonald from whom I learned a great deal.

The experimental animal we used for most of our investigations was the sheep, and although I never became a good animal surgeon I did eventually acquire enough surgical skill to get my animal preparations to work. As a single experiment usually ran over several days and as readings often had to be taken two or three times a day, Alexandra and I would sometimes come in together late in the evening to attend to some half-shorn sheep staring vacantly at us from a home-made pen. I think Alexandra must on such occasions have asked herself where her marriage was taking her, but there was never a word of discouragement either then or in the more difficult times that were to follow. During the course of the year I produced two short reports on the changes that took place in muscle and in the brain when the kidneys failed, and I wrote a more respectable paper on the chemistry of the blood in this condition. But with Oxford in mind I continued to regard my stay in Melbourne and my work there as a passing phase.

My well-laid plans were given a jolt, for the second time, by Panzy. It happened in a totally unexpected way. As a result of reading Charles Oberling's book *The Riddle of Cancer*, which I had picked up in my usual haphazard way in the University Science Library, I had become interested in tumours and was busy reading myself into the subject. When Panzy got to hear of this he called me into his study and asked me, with a gleam in his eye, whether I had ever thought of making my career in cancer research. It transpired that there was a scheme afoot to establish a small cancer research institute in Melbourne, and what Panzy was suggesting was that, instead of going to Oxford, I should spend a couple of years abroad in one of the major cancer research institutes and then return to take up a post at the new institute in Melbourne.

My interest in tumours was at that particular moment very strong, and Panzy's suggestion seemed most attractive. To get things going he asked me to prepare a research programme, and this led to my first encounter with F. M. Burnet. (He became Sir Macfarlane while I was in Melbourne, which elicited some wry comments from the radicals.) It appears that, like Panzy, Burnet was a member of the steering committee for the proposed institute, and Panzy hoped to enlist his support for this new scheme that had me at its centre. The research programme I prepared was sent to Burnet; and fell absolutely flat. Panzy arranged a lunch for the three of us during which it transpired that Burnet thought nothing of the idea that viruses might be involved in producing cancer, and he regarded my immunological suggestions as totally impracticable. He was certainly wrong about the former and I don't think he was altogether right about the latter; but his prestige was such that his opposition alone was enough to put paid to the scheme. My sights returned to Oxford.

On Florey's recommendation I had been awarded a travelling scholarship of the Australian National University, and I was expected to arrive in Oxford in the autumn of 1952. In the meantime Mrs McConnan had returned from her European tour and we had to find somewhere else to live. This brought us face to face with the realities of the furnished accommodation racket. We looked at one derelict place after another and were amazed when we eventually found a couple of elegantly furnished rooms in fashionable Punt Road at a price that we could afford. We hadn't been there long before we realized why the rooms were vacant. The landlady was a prey to hallucinations and, after monitoring every move we made in our two rooms, began to accuse us of all sorts of imaginary transgressions. We fled to a slummy subdivided house in East St Kilda where we were treated through a flimsy partition wall to the intimate telephone conversations of unsavoury neighbours. From there we moved to our last abode in Melbourne, an eccentric but actually self-contained flat in a large neo-Gothic mansion in Hawthorn. After our two previous experiences this flat was a huge relief, and once we had become accustomed to the assortment of aged items that served as furniture, we quite enjoyed living there. The *pièce de résistance* was a small organ that incorporated a pianola device. From this I occasionally attempted to extract Bach. We were not however destined to stay long in Hawthorn.

This time our plans were changed by an entirely new consideration, the advent of what a few months later was to become our son Paul. This development meant that we would either have to leave for Oxford earlier than we intended, so that Alexandra could have the baby there, or we would have to postpone the journey until the baby was old enough to travel. We decided that the first course was preferable, and I wrote to Florey asking whether it would be acceptable for me to arrive in the spring of 1952 instead of the autumn. His reply was not actually brimming with enthusiasm but he agreed to the change of plan. So we began at once to explore what sea passages were available in the early part of 1952. Ships to and from Europe were then very crowded, with long waiting lists, and it was only with great difficulty that we finally managed to obtain berths on the P and O *Stratheden*, which was due to sail from Melbourne at the beginning of March.

Panzy threw a farewell party for us, not as boisterous as the one that saw me off from the Royal Prince Alfred Hospital but heart-warming none the less. As a parting gift, I was presented with a copy of *The Oxford Atlas of the World* in which all the staff had inscribed their names. I use it still, though the names and boundaries of so many countries have changed beyond recognition. We were seen on board by a small group of friends. The departure of a passenger ship from an Australian port has always been an emotional event. Those on deck and those left behind on the pier are joined together by bright paper streamers, held tightly till they are broken by the movement of the ship. 'Auld Lang Syne' is sung until the two groups can no longer hear each other, and hands wave farewell until they can no longer be seen. Before the coming of the aeroplane, if you left Australia at all you left it for a very long time.

The *Stratheden* was my first experience of England. The P and O boats were then still run with almost military precision, and you could see in the behaviour of officers and crew the marks of a long maritime tradition. I was at once fascinated by English speech, kept an ear cocked for interesting regional inflections, and was amazed at the infinite variety of obviously class-based overtones. Many of the stewards were Goanese, and the atmosphere on board was still heavily charged with the attitudes and the diction of the Raj. There was, for example, tiffin, which always offered a range of Indian curries (Khandahar, Johore, etc.), and at sundown you sat on deck in a large cane armchair and drank gin slings or rum punches or whatever took your fancy. We stopped for a day at Adelaide, which I thought a dull place with too many churches, and then crossed the Great Australian Bight to Fremantle, the port for Perth. Perth was then not much more than an overgrown country town, the principal tourist attractions being the University, a pretty collection of buildings on a hill overlooking the Swan river, and a shopping mall that purported to be an exact replica of Old England. We got back to Fremantle only just in time to scramble on board and watch the streamers break for the last time. It was not as moving as I had expected it to be, perhaps because for me the streamers were already broken. But I could not help wondering, as the last shadows of the Australian coast receded, whether we would ever come back.

4

THE YOUNG AUSTRALIAN AT OXFORD

The train sped down from Paddington through fields flecked with patches of snow. April had come but the trees were not yet in leaf and the landscape under a leaden sky still had the aspect of winter. Alexandra sat beside me and gazed, a little morosely I thought, through the grubby window. Having visited England twice since the War she had a more realistic vision of what lay in store for us than I had. Although the War had ended more than six years previously, recovery in England had been very slow. London was still pock-marked with bomb craters, food was rationed, almost everything one needed was hard to get, and there was a severe housing shortage. I viewed all this with perfect equanimity, indeed as something of an adventure. I could hardly believe that the problem of finding accommodation could be any worse in Oxford than it had been for us in the latter part of our stay in Melbourne, and as for the rest, it would only be for a couple of years anyway, or so we thought.

Oxford Railway Station was then and still is a very depressing introduction to the city. In 1952 the Victorian structure lacked the shoddy additions of later days but it had an intrinsic shoddiness of its own. You could indeed see the celebrated spires as the train pulled into the platform, but what really hit your eye was the dispirited provincial grime of the station itself. We were met by a school-friend of mine, a biochemist who had left Australia for Oxford a year earlier and to whom I had written. He had in his hand a few replies to an advertisement he had inserted on our behalf in the *Oxford Times* seeking accommodation, and he had booked a room for us in a boarding-house known simply as 100 Banbury Road, at that time the Ellis Island for overseas scholars arriving in Oxford. We were to live there until we found a habitable flat that we could afford on the far from princely stipend that my scholarship provided.

The following morning I set out to let Florey know that I had arrived: Mahomet had come to the mountain. It was a brighter day with what I took to be a touch of spring in the air and I noticed that the trees, all unfamiliar, had buds on them. As I turned into

South Parks Road, where all the University laboratories then were, it struck me how remarkably asymmetrical the street was. On one side the large, serious, functional structures of the University science departments, on the other the elaborate grotesquerie of Rhodes House on the corner and then an unbroken row of late-Victorian houses. It was as if the street formed a boundary between two immiscible fluids. When I knew Oxford a little better I learned that this dichotomy was at the heart of the University itself.

The Sir William Dunn School of Pathology is at the very bottom of South Parks Road, next to the University parks. Unlike all the other laboratories it is set well back from the building line, behind a formal garden closely planted with shrubs and ornamental trees. Very few laboratory buildings are actually beautiful, but the Dunn School is. Built in 1926 in an unaffected Queen Anne Style it has a gentle air, the deep-red brick set off by a pair of curved, rather ornate, stone staircases leading up to the entrance. One might expect it to be an art gallery rather than a laboratory. The garden beds are screened from the road by a low brick wall and they come as something of a surprise as you enter the grounds. As I walked through the large, wrought-iron gateway that April morning, I was met by a sight so unexpected, and to my innocent eyes so extraordinary, that I stopped for a moment or two and gaped. Beneath the trees and shrubs, the beds were covered to saturation with a carpet of small soft blue flowers with white centres, almost fluorescent in the morning light. I later learned that their botanical name was *Chionodoxa*, the Glory of the Snow, and I do not think that I have ever seen garden beds produce anything more beautiful. They are still there, and their shy reappearance each spring commemorates for me the moment of my first appearance in the sanctuary in which, with one brief interlude, I was to spend the rest of my working life (Plate 4).

The interior of the Dunn School does not disappoint. A splendid oak staircase forms the centre-piece, and the long corridors are punctuated by graceful archways. Visitors are often struck by the height of the ceilings, and I have sometimes wondered whether this might not have contributed in some small way to the uniquely independent scientific style of the place. The laboratory doors, like the central staircase, are of mature oak set off by gleaming brass door-handles. The Professor's study is at the very end of the top corridor, and as I turned into it I saw why the door-handles gleamed

Plate 4. The Sir William Dunn School of Pathology in the spring. The *Chionodoxas* have already flowered, but the garden beds are full of tulips and the cherries are out.

so brightly, for there, halfway along, was a little old man busily polishing them. I presented myself to the Professor's secretary, Miss Winifred Poynton, whose room was adjacent to the Professor's study. Miss Poynton, never known as anything else, looked and was formidable: grey hair combed discreetly back into a bun; neat conventional clothes; self-consciously upper-class in speech and manner. No one ever achieved any degree of familiarity with her or succeeded in prizing out of her any piece of information worth

Plate 5. The Professor's study in the Dunn School.

having. Could the Professor see me? She would enquire. Yes, he could. And so, for the first time, I was shown into the Professor's study.

In the days when the Dunn School was built Oxford expected its professors to be scholars, not administrators, and the architect provided the professor with a study, not an office. It was a spacious, graceful room, the walls lined with oak panels and bookshelves, a fire-place with a marble surround in one corner and in the centre an immense oak desk. Tall French windows opened on to a small balcony, and beyond them I could see the University parks through a screen of bare branches (Plate 5). Florey had not changed much since our first meeting in Sydney, perhaps a little greyer and a little heavier than I remembered (Plate 6). He was even more laconic in his home terrain than he had been then. 'Ah there, Harris, there you are', and pointing to one of the two brown leather armchairs on either side of the French window, 'take the weight off your feet'. Apocalyptic greeting. He collapsed into the other armchair and then asked some desultory questions about our trip to England. I made some desultory replies. Riveting conversation.

Plate 6. Florey at the time of my arrival at the Dunn School.

When I thought that the appropriate moment had come, I asked him what kind of work I would be doing in Oxford. I knew that it was his practice to require all prospective Ph.D. students who had not been through the Oxford system to take the Final Honour School of Animal Physiology before beginning their research. I hoped I might avoid this. I had read pretty deeply in some branches of physiology and had been engaged in physiological experiments during the whole of my stay in Panzy Wright's Department. I did not relish the prospect of returning to undergraduate exercises however informative they might be. Fortunately Florey took the view that I had had enough of an introduction to physiology to make a return to the Final Honour School unnecessary in my case; but he was not yet ready to discuss an experimental programme with me. I gave him a manuscript based on some experiments I had done in Melbourne, in the hope that he would find the work interesting and perhaps encourage me to pursue it. He took the folder and laid it unopened on a small table beside his armchair. 'Harris', he said, 'have you found somewhere to live yet?' 'No'. 'Well, that will be your first piece of research in Oxford'. Lifting himself a little laboriously out of his armchair, he brought the interview to a close: 'Come and see me again when you've settled in'.

I walked back along the corridor where the little old man was still polishing the brass and wondered whether Florey was always so deflating or whether he had made a special effort for me. In any case he was certainly right about the urgency of finding somewhere to live. I took the bus back to 100 Banbury Road, picked up Alexandra, and set about exploring the possibilities offered by the letters we had been given when we arrived at Oxford Station. The first offerings that we saw confirmed Alexandra's worst fears. There had been very little domestic building in Oxford since the War, and in 1952 there were, as far as I can remember, only four purpose-built apartment houses in the city—Belsyre Court in the centre, Bardwell Court, Woodstock Close, and Sollershot in the north—all far beyond our means even if a vacant flat had been available. Neither the University nor the individual colleges took much cognizance of married graduate students in those days, and the only accommodation available to scholars in my position was sets of furnished rooms let at the maximum going-rate by legendary figures known collectively as The North Oxford Landlady.

These rooms were generally to be found in rather neglected Victorian houses, and our first encounters were with the dank basements of such abodes. They were of course dilapidated and usually dirty, the obviously inefficient coal-fires lending their own patina to the overall grime; and the ramshackle bathrooms and kitchens were more primitive than anything we had experienced even in our most difficult days in Melbourne. The prospect of attempting to care for a newborn baby under these conditions was more than we could face, and I began to wonder whether the decision to come to Oxford might not have been a mistake.

Salvation came from an unexpected quarter. One of the addresses we had been given was 18 Beech Road, Headington, but we had been advised against this as it was said to be too far out of town to be convenient. But we went to have a look at it none the less for, by the standards of the sprawling cities of Australia, Headington was barely a stone's throw from the University laboratories. Beech Road is a cul-de-sac terminating in what was then the Headington United football ground, and number eighteen was a conventional semi-detached house rendered in pebble dash. It was neatly painted and looked well-maintained, the small front garden clearly the object of some attention. The door was opened by an elderly lady to whom we introduced ourselves and who in reply informed us in the most undiluted of Oxfordshire accents that she was Nellie Alden. We were taken upstairs to see the rooms that she had to let and were astonished to find two clean, well-furnished rooms and, by Oxford standards, a modern kitchen and bathroom. The rent was five pounds a week, which was four-tenths of my stipend, but we decided on the spot that we would take the rooms if Nellie Alden would let us have them.

There was, however, the question of the baby. It was well known that a baby and, even worse, the prospect of a baby was a serious handicap in the search for accommodation, and it was with great care that I broached the subject. She did not herself mind in the least, but she would have to discuss the matter with her sister Priscilla who shared the house with her and whose bedroom was upstairs adjacent to the rooms that were to let. Familiar with the practices of landlords in Melbourne, we assumed that this was a gentle device for turning us down. It was arranged that she would ring us at 100 Banbury Road that evening to let us have her decision, but we did not really have much hope that we would hear from her again. But the phone did ring for us at 100 Banbury Road that evening and the flat Oxfordshire voice told us that Nellie had spoken to Prissy and that both Nellie and Prissy would be very happy to have us and were especially looking forward to the baby, provided of course that it was not actually going to be delivered in the house.

Easter was almost upon us, and since our rooms in Beech Road would not be ready for a few days and the Dunn School was closed over the Easter holidays, we decided to make a short trip to Belgium, where Alexandra had relatives and close friends. She had lived there for more than a decade, including the years of the German occupation when her survival had hung tenuously on an assumed name and false identity papers. A distant cousin had invited us to spend a few days at a country home that they had in the Condroz, and there I met several members of Alexandra's extended family. One afternoon I was driven over to the nearby town of Huy to see the citadel in which

Alexandra's father had been held captive by the Germans for a year before being transferred to a prison camp in Germany where, against all the odds, he had managed to survive the War. On our way back to London we spent a couple of days in Brussels, a city which we later had many occasions to visit and which has a special place in my evolution as a European. In Australia no building still standing antedates the nineteenth century, and any surviving piece of Victorian Gothic is a monument. Brussels was the first European metropolis that I had had a chance to explore, and it was there that mediaeval streets and ancient buildings first cast their spell. Alexandra of course knew the city intimately, and it was under her expert guidance that I was initiated into its mysteries. The spell proved to be irreversible; for mediaeval towns and the people who lived in them have remained for me a life-long source of fascination.

When we got back to Oxford we found that in our absence spring had unmistakably arrived. The Banbury Road was splashed with the pink and white of flowering cherries and the trees were covered in young leaves. We moved into 18 Beech Road where Nellie was helpful but unobtrusive; and on the first Monday morning after the Easter holidays I took the No. 2 bus down to St Giles and made my way to the Dunn School to see Florey again. This time he was ready to talk about what I was to do. I do not know whether he had read the manuscript I had left with him, but he did not in any case want me to pursue that line of work. His view was that there was no point in my coming to the Dunn School to work on a problem with which none of the staff in the Department had any familiarity: the whole object of the exercise was that I should be exposed to the skill and experience that was available there. This seemed perfectly reasonable and I waited to hear what he had in mind. He then introduced the subject of inflammation, in which he had had a life-long interest, and in particular the role of chemotaxis in this process.

Chemotaxis is a term used to describe the change in direction that can be imposed on the movement of certain cells by substances present in the environment. The phenomenon was first clearly observed in the nineteenth century in studies on bracken sperms which were found to be attracted by substances liberated by the ovary of the plant. Many other examples exist in nature of cells changing the direction of their movement in response to chemical stimuli, and the special interest of this phenomenon in the context of inflammation is that the white cells of the blood, the leucocytes, also show such responses. When the tissues become inflamed, the white cells leave the blood vessels and wriggle about through the tissue spaces where they ingest and destroy any bacteria or other harmful agents that they may come across. The leucocytes are aided in this task by the fact that they may react chemotactically to substances given off by the bacteria. This has the consequence that the leucocytes find the bacteria more rapidly than would be the case if contacts between the two were random.

In 1952 the exact role of chemotaxis in the inflammatory response was surrounded by a great deal of obscurity. Of the three major kinds of leucocyte in the blood, chemotactic responses had been clearly demonstrated in only one. None of the chemotactic substances released by bacteria had been defined in chemical terms, and there was some evidence that chemotaxis could also be exerted by substances released from damaged tissues. One such piece of evidence that attracted a great deal of attention at that time was a claim made by an American pathologist named Valy Menkin that he had purified the chemotactic substance released by damaged tissues: he contended that this substance was a product of the decomposition of tissue proteins and he named it 'leukotaxine' to indicate that it attracted leucocytes. It was leukotaxine that Florey was especially interested in.

Florey's experimental outlook in the early 1950s was naturally still dominated by the clinical triumph that had resulted from the extraction and purification of penicillin in the Dunn School; and it was clear to me from his conversation that his interest in leukotaxine was fed by the possibility that this substance, when purified, might also have some therapeutic applications. I later learned that all Florey's research interests had some therapeutic goal in view. Although in retrospective accounts of his own work Florey often protested that his experiments were undertaken for their intrinsic biological interest and not with a view to therapeutic application, a glance at his publications shows that nothing he did was far removed from some major human disease. Indeed, when I got to know him better I was astonished to find that almost all his work was concerned, in one way or another, with illnesses from which he himself or close members of his family actually suffered. Although I had not read Menkin's book on the subject, I knew a little about leukotaxine and was a good deal less enthusiastic about its prospects than Florey appeared to be. But of course I agreed to have a go at it. There was no option.

There were a number of young Australians in the department at the time, some of them like myself supported by scholarships from the Australian National University. Florey introduced me to one of them, George Mackaness, with whom I was to share a laboratory for the next couple of years. This laboratory, Room 47, was almost next door to Florey's study, so that George and I were very well-placed to observe the often fascinating comings and goings that took place there. Florey asked George to show me round the Department and then disappeared. Two months elapsed before I had an occasion to speak to him again.

The top floor of the department was Florey's personal fief: everyone in it was engaged in some investigation that had been initiated by him and in which he had a direct interest. He had stopped working on antibiotics. His swan song in this field had been the massive two-volume work on the subject produced under his direction by the members of the penicillin team. These volumes remain a repository of interesting historical detail, but by the time the Oxford University Press had actually produced them they were already seriously out of date. The last of his Ph.D. students to work on an antibiotic was J. L. Gowans, later to become the Secretary (Chief Executive Officer) of the Medical Research Council, who studied the properties of an unsuccessful compound called nisin. After that Florey returned to his pre-War interests: inflammation, the physiology of the gastro-intestinal tract, and in particular the properties of mucous membranes. It was almost as if penicillin had been a parenthesis in his life.

The other Australians on the top floor were George Watson and Leigh Dodson, who, unlike George Mackaness and myself, eventually returned to complete their careers in Australia. Three of the penicillin team also had laboratories on the top floor: Norman Heatley, who had played a crucial and, in my view, hopelessly underestimated role in the Oxford work on penicillin, Gordon Sanders, and Margaret Jennings, who later became the second Lady Florey. On the floor below, Edward Abraham and Guy Newton manned the last outpost of research on antibiotics. They were interested in the chemical structure and biosynthesis of the penicillins, subjects that lay outside Florey's range. Next door to Abraham and Newton there was Gareth Gladstone, the Reader in Bacteriology, whose main interest was anthrax; and, at the other end of the corridor, two unforgettable and quintessentially English characters, Arthur Quinton Wells and Sir Paul Fildes.

Wells was the only representative in the Dunn School of a famous tradition in English science, the independently wealthy amateur. He lived in great style in Shipton Manor and was in the year of my arrival the High Sheriff of Oxfordshire. Because of this he was summoned to attend the coronation of Queen Elizabeth II in Westminster Abbey, an honour that, as far as I know, fell to no one else in South Parks Road. He was tall with grey hair combed neatly across the top of his head and a guardee moustache, very public school in speech and manner; and he really did regard science as something of a game. His research, never very intense, was in the field of tuberculosis. Together with a younger associate, John Wylie, he was then engaged in testing the efficacy of a vaccine made with an attenuated strain of tubercle bacillus that infected voles. There was another, highly virulent strain called the Branch strain which was used in the Department for other experimental purposes. This had been isolated from Wells's gardener whose name was Branch. Wells was my first encounter with English affluence in the county style.

Fildes was something quite different. Son of a celebrated Victorian painter, Luke Fildes, he had been born into a cultivated and artistic London environment. He was close to seventy when I first met him, a great bald head surmounting a small trim frame, straight as a ramrod and full of energy. Immensely distinguished, he was one of the founding fathers of chemical microbiology and had come to the Dunn School in 1949 after retiring from his position at the Lister Institute. In collaboration with a young colleague, Desmond Kay, he had embarked in his retirement on what was for him a completely novel programme of work on bacterial viruses which he would discuss with almost boyish enthusiasm. He was unrepentantly, almost flamboyantly, Victorian in his personal style and in his attitudes, and I felt from the moment I met him that I had come face to face with a piece of history. It was a piece of history for which I later developed a great affection.

Most of the basement of the Dunn School was taken up by workshops and general services, but it also contained one laboratory. This was occupied by Kits Van Heyningen who worked on bacterial toxins. Originally from South Africa, Van Heyningen had become a staunch English liberal, and it struck me that in many respects he was an accurate antithesis of Fildes. Lady Florey (Ethel), who was then writing a book on the clinical application of antibiotics, had an office in the basement, but we did not meet until later. George Mackaness finished his tour of the Department by introducing me to two immediately attractive personalities who were destined to play substantial and continuous roles in my life: Peggy Turner and Jim Kent.

Peggy Turner ran the Department's administrative and secretarial services, and although she had not yet been elevated to the position of Departmental Administrator, that was effectively what she was. She was remarkably generous and helpful to newcomers and was soon on friendly terms with them. She was also an inexhaustible source of inside information that nearly always turned out to be accurate. Jim Kent was Florey's personal technician. Florey had taken him on at the age of thirteen when circumstances had forced him to look for work. He had followed Florey from Cambridge to Sheffield and from Sheffield to Oxford, and had been closely involved in all Florey's experimental work. When I arrived at the Dunn School, however, he was rather at a loose end. Florey was not by then doing a great deal of laboratory work with his own hands, and Kent used to attach himself to any experimental programme that looked as if he could make a contribution to it. In this way he helped generations of Ph.D. students and visiting workers and eventually became something of a legend.

If you want to work for a Ph.D. (read for a D.Phil.) in Oxford you must be accepted by and become a member of a college. Florey's Chair was attached to Lincoln College and on his recommendation most of his D.Phil. students were taken on there. In 1952 Lincoln was much smaller than it is now and led a rather obscure existence around its undramatic quadrangles in the Turl. There I presented myself one morning to fill in some forms and be taken by the Dean of Degrees to matriculate in the University. The first thing I learned was that my Australian degrees were not recognized by the University of Oxford. Having become rather accustomed to being known as Dr Harris, I was a little taken aback to find myself addressed once again as Mr Harris and classified simply as an advanced student. This was very different from the practice at the Dunn School where all the young doctors were given their customary titles, and, however much the astringent style that Florey imposed might batter their egos, they were never made to feel that they were back at school. That was precisely how I felt at Lincoln when I was obliged to don what I regarded as a ridiculous black gown and join the column of recently hatched schoolboys who straggled over to the Sheldonian Theatre behind the tutelary figure of the Dean of Degrees. The matriculation ceremony was very brief and was despatched in a completely perfunctory way by some deputy of the Vice-Chancellor. When it was over we were simply dismissed, and it didn't make matters any better to be told that I was now *in statu pupillari*.

Oxford colleges at that time simply didn't know what to do with post-graduate scholars unless they played games. One was indeed granted membership of the junior common room and had the right to dine in Hall, but it was very difficult for an Australian, married and twenty-seven years of age, to make common cause with English schoolboys ten years his junior. Florey, who received a constant stream of overseas scholars, was very conscious of the difficulty, and it was through his efforts that some years later Lincoln became the first Oxford college to establish a middle common room. During the whole of my time as a D.Phil. student I was asked to only two social functions at Lincoln: a sherry party for overseas scholars at the Rector's lodgings and a private lunch that Florey arranged there on the occasion of a visit to Oxford by Douglas Copland, the Vice-Chancellor of the Australian National University.

So I settled down to the problem of chemotaxis. A week's reading convinced me that almost everything that had been written about the chemotactic responses of leucocytes was suspect. Indeed, if it had not been for the existence of an old motion picture made by Comandon at the Institut Pasteur, I should have doubted whether the phenomenon existed at all. The trouble was that in almost all the experiments that had been described, a chemotactic response was simply inferred from the fact that the leucocytes accumulated at some particular site, either in the body itself or under experimental conditions in which the leucocytes were studied after they had been removed from the body. A few preliminary passes with the methods then current convinced me that leucocytes might accumulate at a particular site for a host of reasons other than chemotaxis and that the conclusions that could be drawn from the use of such methods were very fragile. What was wanted was some reliable and rapid method of recording the actual movement of the cells. One could of course make motion films as Comandon had done, but this was very time-consuming and altogether unsuited to the screening of large numbers of compounds, as might be necessary if chemical fractionation of active components was envisaged. I spent a good deal of time wandering rather aimlessly around the Department in search of inspiration.

It came one morning by courtesy of *Life* magazine. I recalled once having seen an issue that contained a splendid photograph of Times Square lit up at night. It must have been taken with a long exposure on a very slow film for the headlights of the passing cars had left a series of bright traces right across the middle of the scene. It slowly percolated into my consciousness that if the leucocytes could be illuminated against a dark background, their movements could perhaps also be recorded as traces on a slow film if only appropriate photographic conditions could be devised. It did not take more than a couple of days to set up the optical equipment needed to try out the idea, and the very first negatives I obtained made it clear that with a little modification the technique could be made to work.

Thanks to Gordon Sanders's interest in microscopy the Dunn School had an excellent collection of exotic lenses, some actually collectors' items made by firms that had long ceased to exist. I was permitted to work my way through these until I found a lens (Busch × 17) and a condenser that gave me just the conditions I wanted. For periods of up to twenty minutes I could record automatically the precise movement of every leucocyte in the field, and the photographs with their brilliant traces against a black background were spectacular. I still find them now, more than thirty years later, reproduced in some of the textbooks, and as visual demonstrations of chemotaxis I don't think they have been surpassed. When I saw those photographs I knew that the main obstacle to a serious investigation of chemotaxis had been overcome, and that the rest of my D.Phil. programme would be by comparison routine (Plate 7).

I set about clattering through it as fast as it would go. They were heady days in Room 47. George Mackaness, who was an extremely skilful experimenter, was doing some marvellous work on the growth of tubercle bacilli within cells, and I was producing a steady stream of highly informative photographs. We were nearly always in high spirits, joked a lot, and felt very superior. Florey had no idea what I was doing. His approach to the supervision of D.Phil. students was that they should be thrown in at the deep end and then allowed to sink or swim. He didn't become interested in them until they had demonstrated that they could swim. One day he came in to see George about something and was about to leave the room when George said to him: 'I think Harris has some photographs you'd like to see'. I took only one out of my pile, but it was the best, and I handed it to Florey without explanation. He looked at it closely

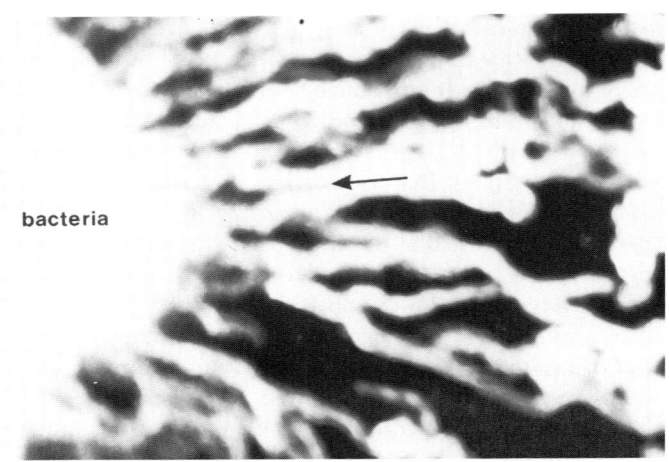

Plate 7. Chemotaxis of white blood cells. This is a record of traces left by the cells as they move in virtually straight lines towards the clump of bacteria on the left.

and of course at once saw the implications. But all he said was that it would have been better if the magnification had been a little greater. And then he walked out. I'm still not sure whether I simply imagined that there was a trace of a smile on his face, but as the door closed George and I burst into peals of laughter.

Although during my first term in Oxford my nose had been kept pretty close to the grindstone, a little of the spirit of the Dunn School and of Oxford itself had managed to seep in. The most notable thing about life in the Dunn School was that it was dominated by research. Nothing else counted for much. Of course the undergraduate teaching had to be done, and was done conscientiously, but no one gained any prestige in the Department simply because he was an able teacher. On the other hand, if someone did an interesting experiment the news went right round the laboratory and at least for a while he couldn't put a foot wrong. This was so very different from what I had experienced in Sydney, where only those who failed in their clinical careers seemed to drift into medical research and where the all-important thing was undergraduate teaching.

It was also very different from what went on in most of the other science departments in Oxford. This difference had its origins in the fact that none of the Dunn School staff were tutorial fellows of colleges. Pathology and bacteriology were taught to the medical students after they had already received their classified B.A.s in the Final Honour School of Animal Physiology and were, as far as Oxford was concerned, graduates. The examination in pathology and bacteriology was unclassified and was regarded by the colleges as an undemanding hurdle to be cleared on the way to the London medical schools where most Oxford medical students went for clinical training. No college thought pathology or bacteriology important enough to warrant the appointment of a tutorial fellow.

In most other departments, for example the Department of Physiology, virtually every member of the permanent staff was a tutorial fellow. This had the result that much (and often most) of his time was spent in undergraduate tutorials and other college activities. Teaching and the performance of pupils in the Final Honour School loomed large in the minds of the physiologists. Research was thus only a part of their lives; at the Dunn School it was the only thing that mattered. Hans Krebs, who was appointed to the Chair of Biochemistry a couple of years after I arrived in Oxford and who had a very exacting view of science, blotted his Oxford copy-book irretrievably by referring to the college dons in his department as part-timers; but as far as their research was concerned there was some truth in what he said. The Dunn School set no store by the system of values that operated in the Oxford colleges. The ability to assimilate information rapidly and reproduce it in well-organized and well-written essays counted for nothing. Wit and verbal brilliance were regarded as social assets but not as academic credentials. There was never any discussion about whether a man did or did not have a first-class mind. What mattered in the Dunn School was not the man's mind but the quality of his experiments. That was the only currency.

All this might sound very drab, but the life of any good laboratory is full of interest, and each one, despite outward similarities, has a different style. The style of the Dunn School was very conservative. Everyone dressed neatly and wore a tie except in the hottest weather. Afternoon tea, that sacred English institution, was taken in the library. It was presided over by Fildes, who would angrily forbid entry to anyone who had failed to change his white laboratory coat for a sober jacket. On a good day Fildes would be full of anecdotes about legendary figures he had met or worked with. He had known Paul Ehrlich, the discoverer of 'Salvarsan', the first chemotherapeutic agent. Salvarsan was an arsenical compound used in the treatment of syphilis, and Fildes claimed that he was the first 'to ram arsenic into the bum of an English aristocrat'. Fildes also held very strong views on the political problems of the day, and the solution that he usually proposed for them was to 'send in the gunboats'. Florey rarely came to tea and when he did his distant manner was rather inhibitory.

In a small room in the basement there was a solitary man who in his way was as much a product of the Victorian age as Fildes. This was Bush, the instrument-maker, who had been separated from the other workshop staff because their inability to meet his standards was a constant source of friction. He was a master craftsman who wouldn't produce anything other than his best and who wouldn't be hurried. For my work on chemotaxis I required at one stage a couple of clips that would keep a plastic slide flat under the microscope. When I asked Bush to make them for me, he questioned me at length about the precise nature of the problem and then convinced me that clips were not what I wanted. In a few days a detailed scale drawing of a slide-holder, to be made in brass, appeared on my desk; and after some weeks Bush presented me with a beautifully machined museum piece which still graces my desk as a paper-weight. A couple of months later, to my amazement, Bush asked my advice about a personal matter. His son was at that time an undergraduate at Oxford reading history, but he was also an excellent cricketer and had been selected to play for the University. Bush's problem was whether he should advise his son to concentrate on history or cricket. I suppose he thought that, as an Australian, I would have a view on this. I advised cricket, and from what I know of the young man's subsequent career that was good advice.

There was another aspect of life in South Parks Road that I found immensely stimulating. This was the fact that I was surrounded by, and could actually talk to, men who bore names that I had read about as an undergraduate. In the Department of Biochemistry there was R. A. Peters, something of a crony of Florey's, very formal in manner but always interested to know what the young men (which meant me too) were up to. He was succeeded by Hans Krebs whom I later got to know very well and about whom I shall have more to say. (I was very flattered to be asked by Florey, while the election to the Chair of Biochemistry was taking place, who the best biochemist in England was. I said Krebs.) In the Dyson Perrins Laboratory there was Robert Robinson, who was held by many to be the most distinguished organic chemist of his time and who had actually once been a professor of chemistry in Sydney. In Physical Chemistry there was C. N. Hinshelwood, in Anatomy W. E. le Gros Clark, in Pharmacology J. H. Burn, and so on. I had the feeling, reinforced each day, that this was where the action was and that I was bang in the middle of it.

However, I did not at all have this feeling about the other famous names that then graced chairs in the humanities at Oxford: Ryle, Syme, Tolkien, Evans-Pritchard, Rhadakrishnan. It is in the colleges of Oxford that the arts and the sciences meet, and since the only access I had to college life was the permission to take meals with the undergraduates at Lincoln, the best I could do was to recognize some of these luminaries from photographs I had seen, if perchance I crossed them in the street or in Basil Blackwell's bookshop. Blackwell's of course soon became a mecca, and still is.

There was thus a great divide between South Parks Road and the traditional collegiate life of the University. You could feel altogether at home in your own laboratory and a welcome guest in any of the others in the science area, but remain a total stranger to the ancient centres of University life. When I arrived in Oxford science was still a problem for the Oxford colleges and it was many years before it was even partly assimilated. One consequence of this was that there was a large number of scientists, some of them quite distinguished, who had no college affiliation at all and lived rather deprived lives as outsiders. None of Florey's penicillin team, for example, had college fellowships until Florey obtained external funds to endow fellowships for them.

The natural meeting place of the outsiders was Halifax House, which had been set up by the University in South Parks Road to provide simple meals and even simpler social amenities for those who had no access to the colleges. It was there that most scientists from overseas had lunch each day and got to know each other. I first met Sydney Brenner, who later became the Director of the Medical Research Council Laboratory of Molecular Biology in Cambridge, at lunch in Halifax House. He was then a D.Phil. student of Hinshelwood's and having a hard time of it. The other social centres were the Lamb and Flag and the Eagle and Child in St Giles, where you could discuss the world at length over a pork pie and a glass of beer.

Outside the laboratories the circles in which we moved were composed very largely of people like ourselves, mainly overseas scholars and the occasional hospitable Oxford man who invited us to his home. There was then a large colony of Australians in Oxford, permanent expatriates as well as transients. Rhodes Scholarships and two or three other similar endowments brought a constant stream of young Australians to Oxford for post-graduate study. The best of them were soon offered positions in Oxford, but they were very rarely offered positions in Australia, or at least not positions compatible with a life of serious scholarship. The end result of this was that, with a few notable exceptions, the most talented Australians lived abroad. It used to be said that, after wool, university professors were Australia's principal export. There were several Australian professors at Oxford when I arrived, some of them so naturalized that no one was any longer conscious of their origins.

There were two very visible features of Oxford University life that as a graduate of Sydney I found very surprising. The first was the formality with which all University business appeared to be transacted. The streets were full of people wearing gowns, short black gowns for undergraduates, longer ones for advanced students, and a variety of more elaborate garments for senior members of the University. In Sydney, gowns were worn only on graduation days and were even then treated as something of a joke. John Anderson thundered against the historic division between town and gown and took the strong view that wearing a gown was a socially divisive act. In Oxford gowns were worn not only at graduation rituals but at lectures, at examinations, even practical examinations, and at every evening meal in college; and they were constantly to be seen flitting in and out of the central University administrative offices. On the great ceremonial occasion of Encaenia the summer streets positively gleamed with the peacock plumes of full academic dress. I came to appreciate in due course that these were the traditions of an ancient university and that Englishmen cherish and preserve their traditions. But I think there is also another, less sophisticated, explanation. Some people love ceremonial for its own sake and actually enjoy getting dressed up. In me, that gene has been completely deleted.

The other feature of Oxford University life that surprised me was the extent to which it was permeated by the religious observances and the ritual of the Church of England. In New South Wales all education supported by the State, from kindergarten to university, was secular by law. There were no school prayers, and the only religious instruction permitted in the secondary schools was an optional period of some forty minutes once a week during which visiting clergymen of the various denominations came to instruct their flocks, if the flocks chose to be so instructed. At the University of Sydney there was no faculty of theology, and only the three or four denominational residential colleges, which housed a very small proportion of the undergraduates, mainly from the country districts, had chaplains. At Oxford I found a flourishing theology faculty and several chairs devoted to various branches of the subject, some tenable only by priests in Anglican orders. Every formal meal in college and many elsewhere began with a grace, and I was told that every meeting of the Hebdomadal Council, the governing body of the university, began with prayers. There was a University Church in which senior academics delivered sermons. Almost every college had its chapel, and the undergraduates, who were obliged to live in college for at least part of their time in Oxford, were committed to the pastoral care of some twenty or so chaplains. With time I came to see that for many members of the University this too was essentially a matter of maintaining traditions, but to begin with it struck me as an astonishing anachronism.

I did not at once fall under the spell of the dreaming spires. Of course I found the ancient buildings and the narrow streets that separated them fascinating, and some of them were indeed very beautiful, but they had not yet won my affection. I think a little of your life has to rub off on to them before buildings become the object of affection; and my life was then being shared between the laboratory and two furnished rooms in Beech Road. The first sign of feelings deeper than admiration came in the early mornings. In some of my experiments records had to be taken at frequent intervals round the clock, and it sometimes happened that I worked through the night into daybreak. There was of course no public transport at that hour, and some years were to go by before I had accumulated enough money to think of owning a motor car. So on those occasions I would walk back home to Headington in the dawn light, either down Longwall beside Magdalen College or, to make a change, through Radcliffe Square and along the High. The streets were completely deserted and the town absolutely still. As I set out, the ancient college buildings would first appear as pale grey masses, mysterious and a little forbidding, but as the summer sun came up, the old stone would break into a smile, and it was surely a smile for me, for I began to think for the first time that it was just conceivable that I could make my life there.

Affection for the English countryside and especially for the gentle rural landscape of the Cotswolds came more quickly. Nellie Alden was responsible for that. She had a Morris Minor motor car which she kept in spanking condition, and from time to time she would shyly offer to drive us out into the country. She knew every corner of the Cotswolds and took us not only to the towns and villages on the tourist circuit but also to out-of-the-way places, a wood full of bluebells, a hill with a long splendid view, an old house tucked away in a cranny. My first reaction to this glorious landscape was not one of surprise but of familiarity. It was almost as if I had often seen it before. And in a sense this was true. For although I had never seen the real thing, my mind had been filled with images of the English countryside from my earliest school-days, and if now and then I exclaimed as some magical vista came into sight, it was not because it was a revelation but because it was just as I had imagined it.

Paul timed his arrival beautifully, in the middle of August when the laboratory was closed. We bought a modest pram which Nellie kindly let us keep in the hall, and it was not long before the young Harrises were to be seen walking their baby along the streets of Headington. Keeping house in those days of food-rationing was an interesting exercise. The Aldens had been butchers in Oxford for generations and they supplied us with our weekly meat ration. We were fortunate in not being great tea-drinkers for this enabled us to barter most of our tea coupons for additional butter coupons. You could, of course, supplement your basic rations in restaurants, but that was a luxury we could afford only rarely. My mother continued the wartime practice of sending us food parcels, and the great durable fruit-cakes continued to arrive long after the need for them had disappeared. We received occasional visits from friends who had been my contemporaries in Sydney, and they were amazed to see me so happy in my new role as paterfamilias. Their amazement was no greater than my own.

The laboratory started up again at the beginning of September and soon I began to see the leaves turn and to understand at last what Shelley meant when he wrote that they were ghosts from an enchanter fleeing. My work was moving fast, and although the subject of leucocyte chemotaxis still did not hold me in thrall, I knew that I was making headway with it. And so did Florey. This showed itself first in the fact that he began to talk to me. Especially in the evenings. His biographers, Gwyn Macfarlane and Trevor Williams, have written about the unhappiness of his domestic life, and the laboratory was full of gossip about it. He used often to dine in Lincoln and, instead of returning to his home in Parks Road, would come back to the laboratory and work quite late. As I often worked late too, our paths sometimes crossed, or he would put his head around the door and strike up a conversation. Not usually about the work, but about things in general. 'How are things?' was often the opening gambit.

I have never met an educated man whose conversation stayed so close to the ground. It was as if he deliberately shunned subtle or polished speech. His accounts of the activities of other people were almost always limited to caricatures of their physical appearances. Someone who had given a lecture was reported as having 'stood up and shuffled his feet'; someone in a state of excitement was 'jumping up and down'; someone else was so frustrated that 'his hair was falling out'. College dons were 'as clever as a bag full of cats'. And although he was obviously interested in what I was doing, his overt reaction was always disparagement. If he came back to Oxford after an absence abroad and I chanced to cross him in the corridor, his greeting would always be 'Ah there, Harris, still going backwards?' or perhaps 'Ah there, Harris, made any discoveries lately?' I naturally found all this cold water rather depressing until I learned by chance that while he was disparaging me to my face he was singing my praises elsewhere. A young pathologist from one of the London medical schools who was spending a couple of days at the Dunn School told me that Florey had recently paid them a visit and had made some very laudatory comments about my work. After that I was perfectly happy to take as much disparagement as Florey was prepared to dish out, but some of the young men working in the Department reacted very differently. Florey's cold style drove them to despair.

You could not talk to Florey for long about science without discovering that he genuinely disliked theory. The scientists whom he admired were those who had done tangible practical things, who had made discoveries that produced visible effects on people's lives. If, in discussing your work with him (a rare event), you began to speculate, his eyes would glaze over and it was clear that he had stopped listening. One morning when he put his nose into our room

I started prattling enthusiastically about a lecture by J. B. S. Haldane that I'd attended the night before. 'Ah yes, Haldane', he said, 'but what has he ever *done*'. About Krebs his comment was: 'I suppose you must have a pretty good memory to devise a cycle as complicated as that'. When I spoke to him about the great excitement being generated by the Watson and Crick model for the structure of DNA, he asked: 'Is that yeast nucleic acid or thymonucleic acid?', terms that had not been used in biochemistry for years. I blotted my copy-book one day when he was talking about Sherrington. I said I thought Sherrington's *Man On His Nature* was hopeless as philosophy and so overwritten as to be barely readable. He froze: 'I wouldn't know about that, Harris, but he was a very good physiologist'. I didn't of course know then that Sherrington had been a major influence in Florey's life and was perhaps the only man that the mature Florey looked up to.

The other side of the coin was that Florey's no-nonsense approach to science inevitably rubbed off on us all. No one who ever worked with him or under his direction remained unmarked by the experience. We all adopted his astringent criteria of what was a good experiment and what wasn't; we all tried to devise simple direct approaches to our problems; we didn't seek to inflate the importance of our work by showmanship or self-advertisement; and when we talked about experiments we told the truth. Some of Florey's young men later became eminent scientists, and people who knew Florey have no difficulty in seeing his mark on all of them.

Although his home-life was a wreck, Florey was hospitable, and one day Alexandra and I received an invitation to Sunday lunch at his home. The old Victorian house in Parks Road in which the Floreys then lived no longer exists. It belonged to the University and was demolished to make way for what is now known as the Keble Triangle development. Florey greatly resented being forced to move ('They're pulling the house down over my head, Harris!'), and he actually lived for many months in his study at the Dunn School while a new but much smaller house was being built for him in Old Marston. To my eye the house in Parks Road was, like all old Victorian houses in North Oxford, rather ramshackle and a little gloomy, but what struck me most was the almost spartan modesty of the furnishings. A glance told you that despite his great eminence Florey was not a wealthy man.

We had a traditional Australian Sunday lunch (lamb and potatoes) which ended with two kinds of cheeses, not the mild and tasty of pre-war Australia but the blue and mousetrap of postwar England. Lady Florey did her best to put us at our ease. She took a personal interest in the young D.Phil. students and made a round of their homes each Christmas, bringing gifts for their children if they had any. However, her hearing was imperfect, even with a hearing aid, and she found it difficult to take part in general conversation. Florey had got into the habit of raising his voice when he spoke to her ('My wife's as deaf as a post, Harris'), and this increased the strain of what was in any case hardly a relaxed occasion.

It was however the first occasion on which Florey discussed Australia with me. We all knew that he was turning over in his mind the possibility that he might return to Australia permanently; but from the very beginning his totally uncharacteristic indecision (and it lasted for years) was linked to his doubts about whether the intellectual climate of Australia could support the kind of academic institution that he envisaged. I formed the impression as the conversation developed that he did actually want to know what I thought, although of course his probing was very oblique; and it wasn't difficult to see that he was sizing me up as a prospective recruit for the John Curtin School which was then beginning to take shape, ever so slowly, in Canberra. As we were about to leave he asked me whether I had read W. K. Hancock's *Australia*. When I said I hadn't, he went over to a bookcase, took the volume out, and handed it to me. 'Bring it back when you've finished with it, Harris'. Hancock's perceptive and totally unsentimental analysis of Australian life was written in 1930, but much of what it had to say was still true when I read it more than twenty years later.

We took our first English winter in our stride. There was for me the novelty of snow which came early that year, but it was only briefly the white wonder I had in my mind's eye, for in the city streets it quickly turned to a dirty grey. Electric radiators kept the two rooms in Beech Road warm and dry, and the Dunn School was actually centrally heated, a rare commodity then in private homes. I still caught the crowded No. 2 bus to work, and the fetid damp that misted up the windows reminded me each morning that the average man in the Oxford omnibus didn't take a bath too frequently.

My work continued to go well through the winter, and by the time the *Chionodoxas* came up again in the Dunn School garden, I had made some inroads into the problem of chemotaxis in leucocytes. I had shown first that all the major types of white cell in the blood exhibited chemotactic responses except one, the lymphocyte. This was a small round cell whose function was then totally unknown. (It was elucidated a few years later at the Dunn School in a classic piece of work by Jim Gowans.) But more important than the fact that chemotaxis could be demonstrated in all the other leucocytes was the finding that they all reacted chemotactically in the same way and in response to the same wide range of bacteria. This was important because one theory put forward to explain why different types of leucocyte accumulated in the tissues when they were infected with different bacteria was that the leucocytes showed differential responses to the bacteria, some being attracted by one kind, others by another. My experiments showed that this could not be the explanation. Chemotaxis, as far as I could see, was a non-specific response that all leucocytes, except the lymphocyte, made when they were exposed to bacteria.

I was not altogether surprised to find that Menkin's 'leukotaxine', the chemotactic substance that he claimed was derived from the breakdown of tissue proteins, was an experimental artefact. A chemotactic substance derived from the breakdown of tissue proteins might well exist, but I was convinced that Menkin hadn't isolated or crystallized it, as he contended. Moreover, with my much better techniques for demonstrating chemotaxis, I was unable to produce chemotactic responses in leucocytes by fragments or extracts of tissues that had been allowed to break down under conditions that excluded bacterial contamination. Florey had reservations about this negative result, but he was sufficiently impressed with the work as a whole to encourage me to write it up and to present it at a meeting in London of the Pathological Society of Great Britain and Ireland. It was my first public performance in England and I took a lot of trouble with it. The results provoked a lively discussion and I greatly enjoyed being the centre of attention. It was at this meeting that I met Alexander Fleming for the first and only time. There was no love lost between Fleming and Florey, but when Florey spotted Fleming on the other side of the room, looking very dapper in a smart bow-tie, he wheeled me over to him and introduced me. Why he did this I don't know. Fleming helped me to a few complimentary remarks and then wandered off.

News of my work had begun to spread around, and one morning I received a surprise visit from Victor Rothschild who was then working in the Zoology Department in Cambridge. He had recently been studying the movement of cattle sperms and took the view that they did not actually show directional responses. He believed that they accumulated at particular sites as a result of a statistically-based redistribution of the organisms, each one moving in an essentially random fashion. This model and the treatment he gave it were very reminiscent of the behaviour of molecules in a gas. An increase in temperature at one point causes a localized increase in the amount of molecular movement, and this produces a redistribution of the molecules away from the region of higher temperature to regions of lower temperature. Instead of molecular movement Rothschild substituted a term for the angle at which the sperms turned (the turning angle), but otherwise his analysis of the problem was not too different from the standard formulation of the gas laws. I thought this approach was a classical case of an investigator imposing preconceived ideas on the data. It seemed to me more plausible, on grounds of evolutionary homology, that if bracken sperms were attracted chemotactically to the bracken ovary, animal sperms could be attracted chemotactically to the animal ovum. Rothschild walked into our room, introduced himself, sat down, and asked me what I was about. I showed him. He could hardly avoid conceding that chemotaxis in leucocytes was a clearly directional response, but he still thought this could hardly be true for cattle sperms. I found him a difficult man to move.

Many years later he became an influential adviser in the Cabinet Office and while there introduced the customer–contractor principle into the administration of British science (the government ministries were the customers, scientists the contractors). I thought the customer–contractor principle was also a preconceived idea imposed on the data, but despite the protests of almost every section of the scientific community he remained unmoved. None the less, when he got back to Cambridge from Oxford he made arrangements for one of his young associates to pay me a visit. This was Charles Brokaw, a young American who was then working in his laboratory and who is now a professor at the California Institute of Technology. Brokaw's visit was a delight, and when he returned to Cambridge he used my method to study chemotaxis in the classical material, bracken sperms. He did some beautiful analytical work on the movement of these creatures and eventually became one of the world's foremost authorities on the mechanisms by which organisms are propelled by cilia and flagella (little hair-like and whip-like structures on their surfaces). What the position is with cattle sperms I still don't know.

While I was busy with chemotaxis, I made one small non-scientific contribution to a totally different subject. Florey turned up one morning with a reprint in his hand. It was written in Italian. He knew about my background in modern languages and asked whether I thought I could translate it into English. The reprint turned out to be Brozzu's original paper describing the discovery of an antibiotic substance produced by a mould of the *Cephalosporium* species. This was the beginning of the work at Oxford that led eventually to the production of a whole new class of antibiotics, the cephalosporins. Ted Abraham, who directed this work, still has my translation. He tells me it's not bad.

I was now in my fifth term at Oxford and was getting rather bored with being *in statu pupillari*. You cannot get a D.Phil. (Oxon) until you have been in residence for six terms, but I was sure I had enough material to make a satisfactory thesis (judging by previous ones I had looked at) and decided to get it out of the way. I did not tell Florey about this but simply got on with the job. I wrote it all out in long-hand and, to save money, Alexandra typed it. When it was finished I presented it to Florey, hoping he would be impressed. He was rather taken aback and didn't seem at all pleased. He said he was rather busy just then but would let me have his opinion when he'd had a chance to read it. A couple of weeks later he returned the typescript without a mark on it but with the comment that it would have been a better thesis if I had waited a bit. He thought none the less that it might just pass muster. At the same time he suggested, rather incongruously, that I should boil it down into a review which he thought might be suitable for *Physiological Reviews*. This could only be interpreted as a compliment, for articles in *Physiological Reviews* were then solicited only from people who had some standing in their field, and to have written such an article meant that you had arrived. Florey also proposed the names of two examiners for my *viva voce* examination, one of whom was Gwyn Macfarlane, who later wrote an account of his early life.

A *viva voce* examination in Oxford is a very stately affair with the examiners and the candidate in *subfusc* : dark suits, white bowties, and, of course, gowns. When mine was over the examiners adhered to the convention of not telling me directly what the outcome was. Examiners can only propose; the Faculty Board disposes.

However, there was nothing in the rules to prevent them dropping a hint, and a few months later I became Dr Harris again, this time Oxon.

When Florey had finished reading my piece for *Physiological Reviews* he thought it was time to talk to me again about what I was going to do next. I was waiting for it. I had already decided that I was not going to continue to work on chemotaxis. I had what I thought were good reasons for this. One was that I was unable to detect chemotactic substances in tissue breakdown products prepared without bacterial contamination, but even if they were there, I judged that it would be very difficult to isolate them with the chemical methods then available. If my judgment was wrong, it none the less remained true that this was a better problem for a chemist than for me. I thought that a much more interesting problem was how cells could sense a chemotactic substance in their environment and, having sensed it, react by changing the direction of their movement. But here too, it was difficult to see what methods one could use to approach the problem. No one had yet isolated a receptor molecule from a cell membrane; indeed the very term 'receptor' was not yet in current use. More than twenty years were to elapse before any serious progress was made in the molecular analysis of the chemotactic response, and this progress depended on the development of chemical and physical methods that simply did not exist when I was working on the problem.

At the heart of the matter, however, was something far less rational. My interest in the cancer problem had not gone to sleep. I continued to read in the subject and became increasingly confident that I could do a good deal better than much of the work I found in the cancer literature. I very much wanted to be given a chance to try. I was intent on studying cell multiplication, and the trouble with leucocytes was that none of them would multiply outside the body, at least not under any conditions that had so far been tried. I needed cells that did multiply after they were removed from the body, so that I could examine what induced their multiplication and what restrained it. My hope was that I could coax Florey into agreeing that this line of work was worth pursuing.

Our meeting was a catastrophe. Florey started off by saying that now that I had devised a good method for studying chemotaxis and had done the groundwork, I should exploit the situation. In particular I should get to the bottom of the leukotaxine question which he still regarded as important. I rehearsed my objections and told him that I thought there was nothing in leukotaxine. He began to get annoyed. To understand why, one must appreciate that for twenty years or more Florey had been the sole and undisputed arbiter in every aspect of the life of the laboratory. His writ ran. Every one of the staff in the Experimental Pathology Division of the Dunn School, whether they were temporary juniors like myself or established seniors of many years standing, was working on a problem of Florey's choosing. No experimental work was done there unless it commanded his continuing interest. And here was I, at the very bottom of the pile, challenging his judgement.

'Well, Harris, what would you propose to do?' So I told him. It didn't go down well at all. To begin with Florey wasn't really interested in cancer. Some years before I came to Oxford, Isaac Berenblum had worked at the Dunn School and while there had discovered co-carcinogenesis, a name which he gave to the observation that skin cancers arise in two experimentally distinguishable stages. However, Florey didn't have such a high opinion of this work: he thought it too theoretical. Neither he nor any other member of the staff of the Dunn School gave lectures on cancer to the undergraduates. These were always given by someone invited from outside, in my time Rupert Willis, who had written an authoritative work on the spread of tumours in the human body, and then Stretton Young from the Imperial Cancer Research Fund Laboratory. In the first edition of Florey's textbook *General Pathology* there are no chapters on tumours, and these were only reluctantly introduced into the second edition because of adverse criticism in the reviews. But, more specifically, Florey thought nothing of the particular approach that I was proposing to adopt. He thought that the problem was much too difficult and that my efforts were doomed to failure.

As he talked my thoughts turned to Alexandra and Paul in our two furnished rooms in Headington and to the easy, affluent, sunlit life offered by clinical practice in Australia; and I decided that if I was going to take the hard and uncertain road of medical research, I would work on a problem of my own choosing and not one dictated to me by someone else, however eminent he might be. When Florey stopped I took a breath and said, as calmly as my excited state would allow: 'Professor, I have decided what I am going to do. All you can decide is whether I am going to do it here or somewhere else'. Florey looked at me as if I had gone off my head, and I looked at him and thought it was high time someone told him where to get off. But I also thought that I had better clear out before I said something totally unforgivable. So I simply mumbled 'I think I'd better go now' and left the room. People will tell you that you can never accurately recall a conversation that took place many years previously. It isn't true. It depends on the conversation.

I went home to Alexandra and told her that we would probably have to leave Oxford. I explained what had happened and, as I expected, met with understanding and loyalty. I thought it was possible that I could get a job in London. Several people there had expressed an interest in me, and since Florey's irascibility was well-known, I thought it might not be held too much against me that I had fallen out with him. If all else failed we would return to Sydney and I would put up a plate as a general practitioner. A week went by, but there was no move from Florey. I wanted to know where I stood so I tried to see him. Miss Poynton said he was too busy. I wrote him a letter asking for an appointment. Another week went by. Then, finally, Miss Poynton came to my room and said that the Professor could see me now. I entered the study once more. Florey, stone-faced, was standing at his desk as I came in. 'You wanted to see me, Harris?' 'I wanted to know whether I have to leave'. There was a pause and then, I thought, that flicker of a smile. 'You need a holiday, Harris. You've been working too hard. You go and take a holiday. I'll see you when you get back'.

So I went up the hill to Alexandra and told her that we were not leaving Oxford after all. I learned many years later that during the icy period when Florey refused to see me he had discussed chemotaxis with Ted Abraham. He trusted Ted's judgement in chemical matters and wanted to have his assessment of the chemical problems that might be involved. It turned out that Ted's views were much the same as mine, and Florey finally accepted that. When I returned from a few days' break I saw Florey again. The meeting was brief and he agreed to let me do as I wanted. As I turned to leave him, however, he added a rider that I have had many occasions to recall in the course of my scientific life: 'But remember, Harris, you're on your own'.

5

ON MY OWN

Perhaps I should offer a more reasoned explanation of why I chose not to play it safe. A dutiful apprenticeship to an acknowledged master looms large in the mythology of science. Hans Krebs once wrote a deliciously unselfconscious article to demonstrate the indispensability of such apprenticeships: he traced a lineage of Nobel prize-winners that terminated in himself. I do not think that I ever had much time for this doctrine. Although most successful scientists are pleased to acknowledge the influence of their mentors and even more pleased to bask in the success of their pupils, there is a good deal of conventional posturing in all this. There are eminent scientists who do indeed generate eminent pupils, but this is not because they teach them some particularly arcane craft, or even some particularly fruitful attitude to science. Eminent scientists naturally attract gifted and ambitious young collaborators, often from all parts of the world, and of these a small number in due course themselves become eminent. However, this is not because of anything special that they have learnt at the feet of the master; it is because they are what they are.

What Hans Krebs failed to see was that in his article he was writing about the genesis of a second eleven, Nobel prize-winners though they might be. (American readers please note: a cricket team has eleven members.) If one looks at a first eleven, for example, Harvey, Newton, Lavoisier, Darwin, Mendel, Pasteur, Koch, Faraday, Planck, Einstein, Ehrlich, then the idea that they were made by the influence of their mentors becomes ludicrous. If they had any mentors at all, they usually rejected their teachings. What distinguishes a first eleven from a second eleven is this. The second eleven work within a framework that is provided for them. The problems that they choose to investigate arise from the work of their mentors, and the approaches that they adopt are, at least initially, recognizably derivative. They make their mark by solving problems that have been posed by previous workers in the field, by devising methods that permit greater accuracy of measurement or novel kinds of measurement, or by extending the methodology of a field with which they are familiar into some related field. These contributions are not to be sneezed at; but they are a far cry from the contributions of a first eleven. The first eleven generate the original conceptual frameworks for whole fields of enquiry, not, of course, out of nothing, but sometimes very nearly. Immense tracts of modern physics have readily discernible origins in Newton; before Pasteur and Koch bacteriology as a science can hardly be said to have existed; before Mendel there was no coherent genetics; before Darwin no clear perception of the principles governing the mutability of the living world; before Ehrlich no rational foundation for chemotherapy. Some of the first eleven did indeed produce disciples; others didn't. None of them conformed. What characterizes their work is the magnitude of the questions they asked and the basic, primary originality of the answers they provided.

Now, conceited as I may have been, I did not imagine when I rebelled against Florey's authority that I was destined to make it into the first eleven. But come what may, I had to tackle a problem that was big enough to fire my imagination and I had to tackle it in my own way. Chemotaxis as a fundamental aspect of cell locomotion would have been big enough, but I couldn't see my way into it; chemotaxis as a minor component of the cellular response to infection was not. And although the approach that Florey advocated—purification of the active substance—had an eminently respectable history, it was, for tissue products, a long hard road that called for faith in the importance of what was being purified. In the case of leukotaxine I didn't have that faith. I do not seek, after all these years, to justify the decision I made. It is possible that if I had done as Florey wanted some interesting science might have come out of it. But for me, then, that orthodox course simply wasn't an option.

Macfarlane's biography of the young Florey is subtitled '*The making of a great scientist*'. If a great scientist is one whose work has the qualities that I have described for my first eleven, then that subtitle is a mistake. For not by any stretch of the imagination could Florey be included in that company. Nor would he ever have viewed himself in that light. But after we had had our disagreement and I had settled down to my own programme of work, I began to see in him qualities that would I think gain him entry into a select company of a different kind. For if he was not, by my criteria, a great scientist, he may none the less have been a great man.

I discovered almost at once that he did not in the least hold my rebellion against me. There was absolutely no subterranean resentment at the flouting of his authority and no desire to see his own view vindicated. He accepted the new situation and waited quite dispassionately to see whether I made anything of my new line of investigation or whether I didn't. I think that if I had floundered he would simply have got rid of me; but when my work began to develop in an interesting way, he was genuinely pleased and willingly gave me any help I needed.

Indeed it seemed to me that he actually came to respect, perhaps even admire, my bid for independence, for, as time went by, our conversations became more frequent and less formal. Sometimes we even laughed together, although none of Florey's colleagues allowed themselves any great degree of familiarity with him, and there was never any question of first names, even for collaborators of more than thirty years' standing. I doubt whether Florey had any close friends as most people understand the term, but I think that, in so far as his nature and the gulf between us permitted, I did win his friendship in the end. Certainly he became the staunchest of my supporters and remained so to the end of his life. Many years later when he was a much mellowed President of the Royal Society, he asked me to dine with him one evening in Queen's College, where he was then the Provost. After the usual excessive dinner that the colleges of Oxford provide on special occasions, he began to reminisce about his long tenure of the Chair of Pathology to which I had by then succeeded him. He assured me that everyone who had come to the Dunn School to work with him had found his way afterwards into a good job. 'I never sacked anyone, Harris', he said. 'Well', said I, 'I can remember one occasion when you came pretty close to it'. His eyes lit up and he chuckled. 'No, Harris', he said, 'not really close'.

I moved down the corridor from Room 47 to Room 50 where there was a little more space which I shared with George Watson, who became a life-long friend. The task I now set myself was to explore in a systematic way the differences between multiplying and non-multiplying cells. The immediate questions were what cells I should use and which of the many differences I was likely to find were the ones that mattered. I at once rejected an alternative approach that was then much in vogue, namely the comparison of cancer cells with what were thought to be their non-cancerous or normal equivalents. The cancer literature was filled with studies of this kind, and an eminent biochemist of those days, Jesse P. Greenstein, had actually written a large book in which thousands of such observations were documented. None of this work was very illuminating, in my view, for two fundamental reasons.

The first was that methods had not yet been devised for growing normal or non-cancerous cells outside the body under the sort of controlled conditions that might conceivably permit the identification of significant variables. Virtually all the studies that had been reported compared cancer cells under one set of conditions with normal cells under another. This made it impossible to decide whether any differences found were attributable to the cancerous state or to the disparity of the conditions under which the observations were made. The second reason was that the whole approach lacked penetration. While differences between cancer cells and their putative normal equivalents continued to pile up, no one appeared able to devise a systematic test to determine the biological significance of the observed differences in the context of the cancer problem. Scientists, after Ernest Rutherford, often describe this kind of activity—the accumulation of data for its own sake—as 'postage-stamp collecting'. It does not enjoy a high reputation.

Now whatever else might be wrong with a cancer cell, one thing that certainly did distinguish it from its normal counterpart was that the cancer cell continued to multiply progressively in the body under conditions in which the corresponding normal cell did not. It therefore seemed reasonable to suppose that if you wanted to know what had gone wrong in a cancer cell with the mechanisms that controlled cell multiplication, you had to understand what these mechanisms were and how they worked. At that time virtually nothing was known about the control of cell multiplication, and it was clear to me that no great progress would be made in understanding the abnormality in cancer cells until we knew a great deal more about the multiplication of normal cells. I thought the most promising approach would be, if one could get the appropriate cell, to compare a normal cell that did not multiply in its natural habitat in the body with another normal cell that did multiply in the same habitat.

The cells I chose were the macrophage and the fibroblast. The macrophage was a cell that I had already worked with in my studies on chemotaxis. It is derived from one of the white cells in the blood and, after leaving the blood stream, its main function in the tissues is to remove foreign material or tissue debris. When the tissues are wounded, macrophages come into the wound site and clear up the mess. They may operate effectively at such a site for many days or even weeks, but they do not multiply there. Methods had been devised for obtaining virtually pure populations of macrophages from the peritoneal cavities of animals, and the cells from this source could be maintained in a healthy state outside the body for periods of up to three weeks, again without any cell multiplication taking place. Controlled quantitative studies on the macrophage thus presented no special problem. Fibroblasts also operate at the site of a wound, but unlike macrophages they do multiply there. Their main function is to synthesize and secrete the products that eventually form scar-tissue. Fibroblasts had been grown outside the body for many years, but never under strictly controlled quantitative conditions. So the first task I set myself was to devise techniques for the culture of the multiplying fibroblasts that were comparable in their precision to the techniques we already had for maintaining the non-multiplying macrophages.

There is in the body a ferment or enzyme called trypsin which is secreted into the intestine and there assists in the digestion of proteins. It breaks the proteins up into smaller units called peptides which are in turn broken down by other enzymes into basic units called amino acids. From these, new proteins are synthesized. When fibroblasts are grown outside the body in culture flasks, they adhere firmly to the floor of the flasks and cannot easily be dislodged. At that time the standard technique for bringing the cells into suspension in order to transfer them to another flask was to scrape them off the floor of the flask with some primitive mechanical tool. However, this was very destructive and yielded unreproducible and variably damaged cell populations. Many years previously, in 1916 to be exact, Peyton Rous and F. S. Jones had tried to use trypsin to detach the cells from the surfaces to which they adhered. (Peyton Rous was a remarkable man who, among other things, was the first to demonstrate that a virus could produce a form of cancer.) The procedure used by Rous and Jones was rather destructive and their cell cultures soon petered out.

It struck me as remarkable that in the ensuing years no attempt had been made to develop and refine their technique; and that is what I set out to do. Trypsin, these days, is used so routinely for the purpose of detaching cells adherent to solid surfaces that it might be difficult to believe that, in the beginning, it was a very tricky business. It took me months to get all the variables under control, but in the end I had a procedure that did enable me to bring the fibroblasts into suspension without loss and to transfer them whenever I wanted from one vessel to another. Using appropriately designed culture chambers, I was eventually able to handle both the fibroblasts and the macrophages with an acceptable degree of precision. (Plate 8).

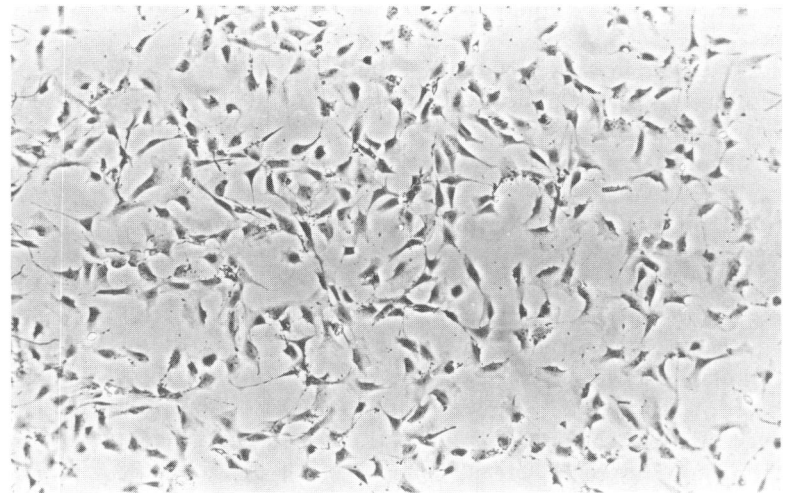

Plate 8. Fibroblasts at last growing in an orderly fashion outside the body.

This modest success came at an appropriate time, for it convinced Florey that I was a good long-term investment. I was in some need of a little goodwill just then, for we were finding it increasingly difficult to make do on the £650 a year that my scholarship provided. To our great regret, we had to leave 18 Beech Road to find some new accommodation that offered a bit more space. This meant, of course, an increase in rent, which I guessed would then consume half of my stipend.

Australian bureaucracy added some delicate touches. The Australian National University took little cognizance of the marital state of its scholars. When I was awarded the scholarship to Oxford, travel funds were provided for me, but if I happened to be married and proposed to bring my wife with me, that was a financial problem that I was expected to solve from my own resources. At one point, an abrupt decision was made to pay the stipends of scholars monthly in arrears instead of monthly in advance, as had been the practice until then. This was no doubt an improvement in accounting practice, but it meant that we had to survive for two months on one month's stipend. As we were not citizens of the United Kingdom, we were not entitled to any child benefits from English sources, but when Alexandra enquired about Australian child benefits, it transpired that we couldn't have these either. The regulations did not permit Australian child benefits to be paid abroad; they were accumulated in Australia and could be claimed when residence was resumed there. I wonder whether three lots of accumulated child benefits still await our return.

When Florey let me know that he was willing to have me stay on at the Dunn School, I was naturally very pleased, but I felt obliged to tell him that I wouldn't be able to make do on the stipend I was then getting. 'How much would you want, Harris?' he asked. 'Twice as much', I said, off the top of my head. That didn't seem to disturb him in the least. He rang Sir Charles Dodds, who was then the Chairman of the Scientific Committee of the British Empire Cancer Campaign, and asked him whether a stipend of £1 300 a year could be provided by the Campaign for a wild but promising young Australian at the Dunn School. It appears that Dodds saw no difficulty. The forms were filled in and rubber-stamped in due course, and I found myself the recipient of £1 300 a year and some research expenses. My only regret is that when Florey asked me how much I wanted, I didn't say 'Three times as much'.

This leads me to a brief digression on the role of patronage in British science. Patronage is now something of a dirty word, and peer review is all the rage; but there is something to be said for patronage if the patrons are the right kind of people. It is a remarkable fact that, despite the small size of the British scientific community and the very small amount of money spent on fundamental research, many of the most important developments in twentieth-century science were initiated in the United Kingdom. In the biology of my own time, with which I am naturally more familiar, I can think at once of half a dozen: penicillin; transplantation immunology; development of methods for determining the sequence of amino acids in proteins; application of physical techniques to unravel their precise three-dimensional configuration; elucidation of the structure of DNA; complete decipherment of individual genes.

In all these cases, behind the scientists who made the advances there stood the patrons. Behind Florey, Mellanby; behind Medawar, Florey; behind Sanger, Chibnall; behind Perutz and Crick, Bragg. The patrons were not at all scientific mentors in the sense described by Krebs. The role of the patron is to recognize an unusual scientific talent and to provide the conditions that make a sustained attack on a difficult and risky problem possible. Not just a salary and laboratory facilities; but a long-term vote of confidence that enables a young scientist to survive repeated disappointment and often years without publication. I think it very unlikely that any of our contemporary peer-review committees would provide long-term support for an investigation that had only a small chance of success, and certainly not if the investigator was a young man with only limited experience. And I can't imagine any such committee continuing its support if, after three or five years, the investigation had yielded no publication.

Yet the distinguished figures who ran British science when I began my career did precisely that. They had great influence with those who distributed the funds, and a strong personal recommendation from one of them could ensure for a young scientist several unharried years in which to try his hand at a really difficult problem. Of course there were many failures, but the remarkable originality of the best of British science in my time was, I believe, at least partly due to the operation of this kind of scientific patronage. I have said enough about Florey to make it clear that he was not my scientific mentor; but he was my patron.

The house we moved to after we left 18 Beech Road was an ugly, late-Edwardian pile in the Iffley Road. Its owner, Mrs Ursula Tyrwhitt, had been a youthful associate of Augustus John's, and the house was full of paintings, some of them her own. She was the exact antithesis of Nellie Alden. Flamboyantly upper-class, she regarded us much as an Edwardian landlord would have regarded his tenants, which, I suppose, was an accurate assessment of the situation. The house was littered with ageing furniture, and in the living-room there was a complete run of *The Yellow Book*. The arrangement proposed was that we should have the front half of the house including the living-room, while Mrs Tyrwhitt would retain the back half, which contained a large artist's studio; but it was understood that she could enter and leave the house by the front door, which involved her passing through the living-room when it suited her. We didn't think this would worry us much, as she was in the habit of spending the whole of the winter in the Canary Islands. It turned out, in fact, that her absence was not an unmitigated blessing, for she left the house in the charge of a housekeeper who monitored our movements and sought to correct them if she disapproved. It was a difficult house to live in, and we hoped that we would not have to stay there long.

Of the numerous theories then being discussed about the biochemical aberrations in cancer cells, perhaps the most interesting was that proposed by Otto Warburg. Warburg was the famous director of a famous laboratory in Berlin. He had been the originator of some of the standard methods of modern biochemistry and before the War had made some remarkable contributions to the elucidation of chemical processes within the cell. His theory was that the fundamental abnormality in the cancer cell was a defect in the mechanism by which the cell obtains energy from nutrient substances. Now the principal nutrient substance for the cells of the body is glucose and they burn it to yield energy in one of two ways. The first and more efficient is called respiration, which involves the consumption of oxygen and the release of carbon dioxide. The second, called fermentation, does not consume oxygen and results in the production of lactic acid, the acid formed in sour milk. Warburg's theory was that while normal body cells derive their energy from respiration, cancer cells have suffered some damage to the respiratory mechanism and are obliged to derive their energy from fermentation even in the presence of ample oxygen. Warburg had put forward this idea in the 1930s and, although it had met with a good deal of opposition, he was still defending it stoutly in the 1950s. Indeed, because of the high technical quality of his experiments and the imperfections of much of the evidence arrayed against him, the question was still unresolved.

I had been through the literature on this subject with great care and, as I expected, was unable to find any fault in Warburg's experiments as exercises in metabolic biochemistry; but I had a profound objection to them on other grounds. In choosing normal cells with which to compare his cancer cells, Warburg was faced with the same problem that had confronted me when I decided that I had to devise a method for handling fibroblasts quantitatively outside the body. He had no difficulty in finding cancer cells: there were some that could be grown as suspensions in the peritoneal cavity of laboratory animals and that continued to multiply in suspension even after they were removed from the body. They were thus ideally suited to the methods he had devised for measuring the consumption of oxygen and the production of carbon dioxide. But when it came to normal cells that multiplied outside the body, there simply weren't any that could be handled in the same way. So, for his prototypal normal cells, Warburg chose sea-urchin eggs. These did multiply in suspension and were thus suited to his methods of measurement. When suspensions of cancer cells were compared with suspensions of sea-urchin eggs, what Warburg claimed was undoubtedly true. But it seemed to me preposterous to regard a sea-urchin's egg as the equivalent of a normal mammalian body cell; and since I now had an adequate method for studying normal body cells that multiplied outside the body, I decided that the first thing I would do with these cells was test Warburg's ideas against this more appropriate material.

However, before this could be done yet another method had to be devised. Warburg's techniques for measuring gas exchanges in cells could not be applied to normal fibroblasts. These techniques were based on the measurement of gas pressures (and were hence known as manometric techniques) and they required continuous shaking of the cell suspensions in order to ensure adequate mixing of the gases. Fibroblasts couldn't be shaken in suspension: if you did that the cells simply clumped together and disintegrated. So I had to work out some way of measuring oxygen consumption in the small culture chambers that I had designed specially for quantitative work with the fibroblasts. I did this by constructing a very small oxygen electrode which could be inserted through an opening into the culture chamber and which, once inside, could monitor the oxygen concentration continuously. An oxygen electrode is essentially a device for measuring oxygen concentration by means of an electrical circuit. In my case, the electrode was a small piece of platinum wire. The oxygen was adsorbed to the surface of the wire and, if appropriate electrical circuitry was connected to it, the amount of current generated was proportional to the amount of oxygen adsorbed, which in turn was proportional to the amount of oxygen present. I got all this to work very well and was soon making measurements on macrophages (non-multiplying) and fibroblasts (multiplying).

The results fell out and I was able to draw two important conclusions from them. First, even in the presence of ample oxygen, both the macrophages and the fibroblasts consumed most of their glucose by fermentation. Warburg's idea that fermentation in the presence of adequate oxygen was the hallmark of a cancer cell was clearly wrong. The second conclusion was more important. I found that even after all the oxygen in my closed chamber had been exhausted, the macrophages remained alive and well and the fibroblasts continued to multiply. This meant that there were normal cells in the body that could operate perfectly well in the absence, or at least in very low concentrations, of oxygen. When one stopped to think about it, this should not have been all that surprising for macrophages and fibroblasts, since, in the healing of a wound, these cells were active even before a normal blood supply, and hence a normal supply of oxygen, was restored to the wound site. It followed from this fact alone that these cells had to be able to operate at very low oxygen levels. However, physiological and biochemical ideas were so dominated by the importance of oxygen for life processes that it came as something of a shock to find that there were cells in the body for which oxygen was not all that important.

I published all this in the *British Journal of Experimental Pathology*, which had been founded by Paul Fildes and with which Florey had had a long association. This was a regular avenue of publication for work at the Dunn School, but it had a serious disadvantage for my kind of work. It was not much read by biochemists, and my findings made less impression than they would perhaps have done if they had been published in one of the standard biochemical journals. If you're ploughing a lonely furrow, it's always a problem to know where you should publish your work.

These experiments brought me two good American friends. The first of these was Alexander Leaf who later became the chief physician at the Massachusetts General Hospital. He was then spending a sabbatical year with Hans Krebs in the Biochemistry Department at Oxford. He was especially interested in the transport of sodium across cell membranes and was studying this process in isolated pieces of frog's bladder. I don't know how he heard about my little gadget for measuring oxygen levels in small volumes, but he turned up in the laboratory one day and asked to see how it worked. Then he set it up in his laboratory in Biochemistry and with it was able to show that the frog's bladder could transport sodium in the absence of oxygen. This too was a surprise for those interested in transport processes.

It was through Alex Leaf that I came to the notice of Hans Krebs who, however, viewed my findings with some reserve. Krebs had been a pupil of Warburg's and had been profoundly influenced by him. Indeed, it was his experience with Warburg that convinced him of the indispensability of a high-class scientific apprenticeship. For his own work Krebs adopted Warburg's patently reliable but limited manometric techniques and stuck to them for the rest of his life. He viewed oxygen electrodes with great suspicion, and he didn't like to see the work of his master so summarily dismissed. In a biography of Warburg that he later wrote, Krebs emphasized that although criticisms could be made of Warburg's ideas about cancer, there was nothing wrong with his experiments. As biochemistry, perhaps so; but as biology they were terrible.

A few years later I had the pleasure of staying at Alex Leaf's home in Boston. I found that by the time I got up in the morning Alex had already been up for a couple of hours working through the mathematical basis of thermodynamics, on which he had agreed to give some lectures. It therefore came as no surprise when in due course he reached the top of the medical tree in Boston. The other American friend I made at that time was William R. Barclay. He had come from the University of Chicago to spend a sabbatical year at the Dunn School, and Florey had suggested that he might like to work with me on the utilization of glucose by cells at low oxygen levels. Bill made the lactic acid measurements, and we published a paper together. He eventually became the Chief Executive Officer of the American Medical Association.

In the laboratory, time went by peacefully, but our life in the house on the Iffley Road was becoming increasingly tedious. Indeed, we were by now overwhelmingly tired of furnished rooms in general, having lived in nothing else since we married. The development of the John Curtin School in Canberra was going very slowly and, as far as we could see, very badly; and even if I were eventually offered a position there it would be years before the laboratories would be ready for occupation. It therefore looked as if we were destined to stay in Oxford for some time yet. Unfurnished accommodation remained unavailable for someone like myself who was neither a fellow of a college nor a member of the University. Although I had no security, I did have the expectation that my annual grant from the British Empire Cancer Campaign would continue, and on the strength of this I began to explore the possibility of obtaining a mortgage with which to buy a small house. It turned out that the combination of my annual grant, a medical degree, and a clean bill of health was enough to procure the mortgage I needed from one of the medical insurance agencies, provided that the total sum borrowed did not exceed two-and-a-half-times my annual salary. I took the plunge and bought a little house for exactly that amount.

It was one of a row of six identical houses then being built in a large field (Rotha Field) at the far end of North Oxford. The houses were joined to each other by their garages and were therefore classed as detached rather than semi-detached, but the difference was a technicality. The new road was called Rotha Field Road and our house was number 4. We bought the furniture from Elliston & Cavell, an old established firm in the city, entirely on credit. It was very simple but well-designed and made of wood. Much of it is still with us. We moved in as soon as the building work was finished, or almost. For those who have not had the experience of a long sequence of furnished lodgings it may be difficult to imagine the pleasure we had in finding ourselves at last in our own home, even if mortgaged to the hilt, with our own sticks of furniture, even if ours only by hire-purchase, and with our own small garden. Suburban domestic tranquility has had a hard time from the makers of the modern novel, but for us Rotha Field Road was a period of great happiness. Our daughter Helen arrived at the end of July 1955 and Ann, the youngest of our children, in the winter of 1956. We received our first visit from my parents while we were living in Rotha Field Road. My father was very disappointed that I had so little to show for all those years of study and urged me to return to Australia and get down to the business of earning what he regarded as a reasonable living. I saw no way of taking his advice even if I had wished to do so. Rotha Field Road remained our home until we left for America in the autumn of 1959.

Having convinced myself that cancer was not caused by a defect in the respiratory mechanisms of the cell, I decided that it might be of interest to have a look at the way in which cancer cells put together their proteins from the basic building blocks, the amino acids. There were some suggestions in the literature that there might be differences between cancer cells and their normal equivalents in this respect, but there had been no systematic investigation of the question and certainly none that met the criteria of precision and comparability that I now regarded as essential. So I set about looking at protein synthesis in my multiplying and non-multiplying normal cells and added a third cell-type which I could handle in the same way, a line of cancer cells called HeLa. (HeLa was the code name that had been given to this cell-type by George Gey who had isolated it from a uterine cancer in a patient called Henrietta Lacks.)

The work involved the use of radioactively labelled amino acids and it was the first time such compounds had been used in the Dunn School. Indeed I think it may well have been the first time that radioactive compounds were used in any of the biological departments in Oxford. We borrowed from the Atomic Energy Research Establishment at Harwell a very primitive device for measuring the radioactivity, and several people from other laboratories in South Parks Road popped in to see how it worked. I studied the incorporation of every one of the amino acids into protein and constructed a detailed pattern of incorporation for each cell-type. A similar sort of study had recently been made on the colon bacillus (*Escherichia coli*) by a group working at the Carnegie Institution of Washington, but nothing of the kind had previously been done with mammalian cells.

However, from my point of view the investigation was something of a disappointment. For, while it was all good useful information, nothing very startling emerged, and certainly nothing that was characteristic of the cancer cell. I did however stumble on one interesting new finding. I discovered that the normal fibroblasts had to have certain forms of sulphur in the medium if they were to survive: without these the cells rapidly disintegrated. This phenomenon later became the subject of intensive investigation, and is still being investigated, but, for reasons that will presently become apparent, I chose not to pursue it. In the work on protein synthesis I was joined by Marianne Jahnz, my first personal technician. The Cancer Campaign had agreed to provide a salary, and when I advertised the job a very serious application arrived from Switzerland. I asked Marianne to come over to Oxford for an interview and then offered her the job. She stayed with me until she left to get married eight years later.

While I was busy with all this, some changes had taken place on the top floor of the Dunn School. George Mackaness had returned to Australia to set up the nucleus of a department of experimental pathology in some prefabricated huts that constituted all there then was of the John Curtin School of Medical Research. His place in Room 47 was taken by John French, freshly returned from a year at the University of Chicago. And George Watson, with whom I had shared Room 50, had left to take up a position in the Radiobiology Division of the Atomic Energy Research Establishment at Harwell. His place was taken by Jim Gowans. Jim, having done his D.Phil. in the Dunn School under Florey's supervision, had gone off to Paris to spend a year or so at the Institut Pasteur. Both John French and Jim Gowans returned to work on problems that Florey set them.

I have already mentioned that essentially all of Florey's research was concerned with diseases from which he or some member of his family suffered. His life-long interest in the physiology and pathology of the intestinal tract arose from problems he had with his own. As a young man at Cambridge he had suffered from severe dyspepsia, which had been misdiagnosed as achlorhydria (absence of the hydrochloric acid that is normally found in the stomach). It appears that for a while he had been advised to take hydrochloric acid with his meals. ('Didn't do anything but rot my teeth, Harris'.) By the time I knew him, he did not seem to have anything wrong with his stomach, but he was often much concerned with his bowels. ('You don't know what a pleasure it can be to make a formed stool, Harris'.) To what extent all this suffering had its origin in nervous tension can now only be a matter of conjecture, but as far as his own bodily functions were concerned Florey was certainly a very introspective man.

When I arrived at the Dunn School in 1952 Florey would usually come in through the main entrance at about 10 a.m. and then trudge slowly up the great oak staircase. A couple of years later there was a sudden change in his habits: instead of walking up the stairs he began to take the small lift to the top floor. When I noticed this I wondered whether he might perhaps be having some more serious problem with his health, for people do not change the habits of a lifetime without good reason. I learnt many years later, from Dr Alan Richards who had been Florey's doctor and was also mine, that Florey had had his first attack of cardiac pain even before I came to Oxford. He had before that been almost a chain-smoker but at once abandoned the habit and began to manage his life carefully. He was very secretive about all this, for he felt that if it were widely known that he had a serious heart condition, he would not be given the opportunity to do things that he was still ambitious to do. No interpretation of the last twenty years of Florey's life can be other than superficial if it does not take into account that during the whole of this period he was subject to recurrent bouts of angina. The emergence of his interest in arteriosclerosis, and quite specifically arteriosclerosis of the coronary arteries, was prompted by his own cardiac pain. John French was the first of a group that Florey now assembled to initiate a programme of research on arteriosclerosis. He was soon joined by Donald Robinson, a biochemist, and John Poole, a specialist in the disorders of blood-clotting. They remained together for several years.

At that time the great debate about the role of dietary fats in the causation of arteriosclerosis had only just begun, and the problem that John French was initially set to investigate was the mechanism by which ingested fats were cleared from the blood stream. Ingested fats, in the form of a fine emulsion, make their way into the blood stream by a channel known as the thoracic duct. Before the War, Florey had worked out a technique for collecting the contents of the thoracic duct in rabbits, but it involved a tricky surgical procedure and was not suited to long-term experiments. In 1948 an American, J. L. Bollman, devised a much better technique that was applicable to rats. This permitted a small tube (a cannula) to be inserted into the thoracic duct and maintained there for many days. John French was the first to cannulate a rat thoracic duct in the Dunn School. He was interested in collecting the emulsion of fats that the duct contained and discarded the suspension of cells that was also there.

It was known that the thoracic duct contained huge numbers of lymphocytes. Lymphocytes constituted about twenty per cent of the white cells in the blood, but essentially nothing was known about their function. Florey had made some abortive attempts in 1940 to determine the function of lymphocytes and was even then very puzzled by their presence in such large numbers in the thoracic duct. By the time Jim Gowans returned to the Dunn School, John French was already cannulating the thoracic ducts of rats by Bollman's technique, and the problem that Florey set Jim was to determine the fate of the cells that John French discarded—the lymphocytes. So Jim set up Bollman's apparatus in Room 50 and began to chase lymphocytes. Five years later he had elucidated their life cycle, and by the time he finally gave up experimental work to become the Secretary of the Medical Research Council, he had made quite fundamental contributions to our understanding of their role in immunological reactions.

I come now to an unhappy and to me still inexplicable episode which, as it turned out, resulted in my never returning to live in Australia. The saga of Florey's relationship with the Australian National University is described in some detail in Trevor Williams's account of the latter part of his life. After a brief period of idealistic enthusiasm, the project became bogged down in a quagmire of administrative incompetence and bureaucratic obstruction. Initially it was assumed that Florey would go out to Canberra when the buildings were up and would assume the Directorship of the John Curtin School. But as time went by, he began to vacillate and at one point actually threw in his hand. However, in the mid 1950s, a change of heart, or of practice, by the authorities in Australia succeeded in resurrecting his interest, and he once again began to talk to colleagues in the Dunn School about prospects in Canberra. He asked me whether I would be willing to go there and on what terms.

Having in mind the past record of the Australian National University and of the bureaucrats that ran it, I said that I would be very willing to go to the John Curtin School if he went out as Director, and my price was a small autonomous department in which I could advance the subject of cell biology. (In the official documents, this proposal appears as a Chair of Cytology, but of course a much broader field was envisaged than the morphological study of chromosomes which was one way in which the word cytology was then used.) Florey thought this eminently reasonable and included a department for me in his plans. However, he was still not altogether prepared to commit himself and suggested that, in the first instance, he and several of his colleagues, including myself, should go to Canberra for a sabbatical year to test the temperature of the water. This idea, as might have been expected, was not greeted with great enthusiasm in Australia, but in the early part of 1957 Hugh Ennor, the Professor of Biochemistry at the John Curtin School and then acting as Dean, was none the less sent over to Oxford to negotiate.

The two days that Ennor spent in Oxford are vividly remembered by all the inhabitants of the top floor of the Dunn School. Florey was closeted with him in his study for hours on end, but it was obvious from the appearance of both men as they emerged from time to time that the negotiations had not gone well. To me Florey muttered: 'I think he's trying to wreck the whole scheme. He's not prepared to negotiate at all'. The most charitable view of Ennor's performance was that he was rigid in carrying out the instructions he had been given. The least charitable view was that he torpedoed the negotiations deliberately in the hope that, if Florey did not return to Australia, he might himself become the Director of the John Curtin School. I must say that at the time I subscribed to the least charitable view. When Ennor came in to see me on the second day of his visit, he informed me, unasked, that if having me in Canberra was the price of getting Florey there, then it was possible, but not certain, that the authorities in Australia might agree to it. But he wanted me to know that nobody out there was at all eager to see me on the site. I have often since wondered what lay behind these endearing phrases. Since I did not know Ennor or anyone else in Canberra at all well, and certainly not well enough to have generated such a high level of personal antagonism, it was difficult to see where such a groundswell of opposition could have come from. It seemed to me more probable, in the light of what Florey was saying, that Ennor's remarks to me were merely an element in his scheme to stop Florey going to Canberra. Be that as it may, the immediate consequence of his visit to Oxford was that both Florey and I washed our hands of the John Curtin School of Medical Research.

Later that year I received a very friendly letter from Panzy Wright. He was a member of the Council of the Australian National University and was aware of the débâcle precipitated by Ennor's visit to Oxford. Indeed, among Florey's papers there is a letter from him to Panzy protesting about Ennor's attitude to me. Panzy now asked whether I would be interested in a position at the Cancer Institute in Melbourne. This was the institute with which I had previously been involved in the abortive scheme that Panzy had devised when I was working in his department. Panzy suggested that I should visit Melbourne and see for myself how things stood. It was a generous offer, but when I discussed it with Florey he advised me strongly against it. His view then was that if you had any dealings with an Australian university, you should take your lawyer with you. He must at some stage have said as much to Panzy for when, with reluctance, I declined Panzy's offer, he wrote back to say that if I had been willing to come out to Melbourne to discuss the proposition, he could have provided me with a good lawyer.

I did not receive another offer of a job in Australia until several years after I had succeeded Florey in the Chair at Oxford. Then, within a few months of each other, I was offered a chair in a proposed new school of biological sciences in Canberra and another in the John Curtin School. Since neither Alexandra nor I had ever been to Canberra, and it was by no means certain that our children would take kindly to life there, I replied in both cases that it was not reasonable to expect me to be able to make a decision of this magnitude without having first spent a little time in Canberra with my family. But that, it transpired, was too much to ask.

My work on the synthesis of proteins in cell cultures brought me into contact with the ideas of Jacques Monod who was a very influential figure in biological research a generation ago. In France his authority had almost papal dimensions. Highly intelligent, elegant, a gifted expositor, and a merciless controversialist, he laid down the law in some of the areas of molecular biology that were then fashionable. I expect it will generally be agreed that the most permanent of his contributions was the work he did with François Jacob on the mechanism of enzyme induction in bacteria; but he was given to expressing strong, even intransigent, views on a wide variety of subjects even beyond the boundaries of biology. He was one of the few experimental biologists I had ever met who set much store by formal logic in the analysis of biological problems, and the scientific arguments he advanced always seemed to me to have a strong Cartesian flavour. And he was a great generalizer. Whatever he found in the bacterium *Escherichia coli*, which was the main object of his investigations, he immediately extrapolated to the living world as a whole. 'What is true for *E. coli*', he claimed, 'is also true for an elephant'.

At the time that I began my experiments on protein synthesis, it was generally assumed, after the work of Rudolf Schoenheimer, that the proteins in the body were in a state of dynamic flux. Schoenheimer was one of the first to use isotopes (elements with marginally different physical characters) to study metabolic processes in the body. He showed that many constituents of the body, including proteins, were constantly being synthesized and broken down, a process known as 'turnover'. His book, *The Dynamic State of Body Constituents*, which appeared in 1942, was a biochemical classic. By 1950 nobody doubted the general validity of Schoenheimer's conclusions. Then, in 1955, a paper appeared from Monod's laboratory describing a study on the turnover of protein in rapidly multiplying bacteria. Monod and his colleagues found that in such cells they were unable to detect any turnover. One perfectly plausible interpretation of this finding might have been that in the multiplying bacteria any breakdown of protein would be very difficult to detect against a background of rapid protein synthesis. However, Monod chose to generalize his finding, and he argued that what he had found in bacteria was also true for the cells of the animal body. He proposed that the proteins within the body cells were completely stable and that the protein turnover described by Schoenheimer was due to cell multiplication and cell death. This was so fundamental a reorientation of our ideas about metabolic processes that no one who, like myself, was involved in the study of protein synthesis, could ignore it. I must say that from the very beginning I found Monod's proposal very difficult to accept.

With the exception of a few rare cancers, none of the cells in the adult animal body multiplied in the unrestrained way characteristic of a bacterial culture, and some of them didn't multiply at all. Moreover, there was nothing like enough cell multiplication and cell death in an organ like the adult liver or adult kidney to account for the turnover of protein observed. It seemed to me that the appropriate material on which to test these ideas was a population of normal non-multiplying cells that could be studied under precise quantitative conditions outside the body. So I turned once again to the macrophages that I could maintain in excellent condition outside the body for many days, and I measured the turnover of protein in them. In this work I was joined by John Watts who was my first D.Phil. student. We soon found that there was indeed a substantial turnover of protein within the cells, and we were able to demonstrate it unequivocally and with precision. So Schoenheimer was right after all, and Monod was wrong. We published our work in *Nature*, then as now a widely read journal, and the idea that the proteins of the cell were completely stable died a natural death.

In 1957 I attended my first international congress. These functions have now become monstrosities to which young scientists should none the less be sent in order that they can experience at first hand the worst aspects of scientific life. But thirty years ago it was still possible to attend an international congress, especially in a minor subject as cell biology then was, and gain some scientific profit from the exercise. The IXth International Congress for Cell Biology was held in the picturesque Scottish university town of St Andrews, and the numbers attending were small enough for one to be able to meet and talk at leisure to anybody one was interested in. I met there a number of attractive people who over the years have done much to enliven my life. But the most important thing that happened to me at St Andrews was that I was exposed to an example of the very best and an example of the very worst kind of science.

The very best was a lecture delivered by Joachim Hämmerling. Hämmerling is one of the great unsung heroes of twentieth-century biology. He was the first to demonstrate that the cell nucleus, which contains the genes, releases into the surrounding parts of the cell, the cytoplasm, substances that determine the cell's development and its specific morphological character. This now seems such a platitude that it is difficult to appreciate that there was a time when the phenomenon had to be demonstrated and that the demonstration met with scepticism and fierce opposition. Hämmerling made his fundamental discoveries long before we had any idea of the chemical structure of genes or of the molecules that carry information from the genes to the rest of the cell.

He worked all his life with a single organism, *Acetabularia*. This is a seaweed with some extraordinary properties. Each plant is a single giant cell, in some species up to several centimetres in length, and it contains only a single nucleus that resides in one of the rootlets. When Hämmerling discovered this, he saw at once that it would be easy to remove the nucleus or to transplant the nucleus of one cell into the cytoplasm of another. He was thus able to examine for the first time what a cell can do without its nucleus and what a nucleus can do when it finds itself in a foreign cytoplasm. Hämmerling's observations on *Acetabularia* cells manipulated in these ways provided the fundamental biological rules that govern the flow of information from the genes to the cytoplasm of the cell.

Although some of this classical work had been done in Germany before the War, it did not become widely known to English-speaking audiences until some of Hämmerling's papers began to appear in English in the 1950s. At St Andrews he gave a talk on his work in one of the smaller auxiliary lecture theatres. His English was imperfect and his delivery dry and halting. But the experiments he described were a conceptual and technical masterpiece; and I can recall very few occasions when I have left a lecture theatre so deeply moved. I later got to know Hämmerling well, and his ideas had a profound influence on the interpretation I gave to some of my own experiments. When he died, I was asked to write a biographical memoir of him for the Royal Society of which he was a foreign member. In that memoir I tried to make restitution for the neglect that had been Hämmerling's lot for most of his life. I wish the *Biographical Memoirs of Fellows of the Royal Society* were more widely read.

The great event of the conference was a plenary lecture in the main auditorium by J. Benoit of Strasbourg. Benoit claimed that he and his colleagues had succeeded in transmitting some of the genetic characteristics of Khaki Campbell ducks to Pekin White ducks by injections of purified genetic material (DNA) isolated from the Khaki Campbells. The claim had been taken up by the press and, of course, hailed as a 'breakthrough'. To be able to change the genetic character of an animal by injections of genetic material from another was, after all, no small claim, and it is hardly surprising that the newspapers made a fuss of it. What was surprising was that the organizers of the Congress took Benoit's claim seriously enough to make his lecture the centre-piece of the meeting.

It was by studying genetic changes produced in bacteria by cell extracts that Oswald Avery, Colin McLeod, and Maclyn McCarty had made the stupendous discovery that genes were made of DNA. But any student of the experiments of Avery and his colleagues could see at once that the claims made by Benoit were wildly improbable. His lecture was a splendid performance punctuated by numerous colour slides of Khaki Campbell and Pekin White ducks; but for anyone with a modest critical sense it was obvious that the whole thing was a piece of monumental self-deception. I happened to sit next to Benoit at lunch the following day and had a second dose of the distinguishing features of Khaki Campbell and Pekin White ducks. Benoit's plenary lecture in the main auditorium and Hämmerling's hesitant exposition in the small lecture room fixed in my mind forever the difference between a 'breakthrough' as identified by the popular press and a real scientific advance.

A further exploration of Monod's views on the stability of cellular constituents led me to make an observation which, when its significance was finally appreciated, turned out to be a real discovery. However, between the observation and the general appreciation of its significance, many years were to elapse, and those that immediately followed the initial publication were the hardest of my scientific life. Some of Monod's colleagues, having convinced themselves that there was no turnover of protein in cells, turned their attention to RNA (ribonucleic acid). In similar experiments with rapidly multiplying cells, they found that there was no turnover of RNA either. Since RNA was known to be the material that carried the instructions from the genes to the cytoplasm of the cell, this conclusion seemed to me implausible. For if the RNA in the cell were completely stable, the genetic instructions, once delivered to the cytoplasm, could not be changed except by a slow process of progressive dilution that took place as the cells multiplied. But this would hardly do for most of the cells in the animal body, and certainly not for those that didn't multiply. So I thought I had better have a look at RNA turnover in macrophages, as I had previously done for protein. John Watts and I made a thorough study of the question and were able to show that there was a substantial turnover of RNA in macrophages, indeed a much greater turnover of RNA than of protein. We published our results in the *Biochemical Journal*, and stability of RNA went the same way as stability of protein.

It was of course of great interest to know where in the cell this substantial turnover of RNA was taking place. Autoradiographic techniques were just then becoming available, and it seemed to me that the application of these techniques might throw some light on the problem. Autoradiography is a procedure that permits the localization of radioactivity within the cell. If a cell is exposed to a radioactively labelled compound that is incorporated into some cellular component, the site at which the radioactive compound is finally located may be revealed by exposing the cell, prepared in an appropriate way, to sensitive photographic film. The radioactivity blackens the film and, under the microscope, its location within the cell is revealed by the presence of black grains in the overlying film. It seemed to me that if I made an autoradiographic study of individual cells in parallel with the biochemical analysis that had revealed the RNA turnover, some further information about the process might well emerge.

Autoradiographic experiments of this kind on RNA labelled by the incorporation of radioactive precursors had been done with one or two other cell-types, and the results obtained conformed to expectations. What were these expectations? Since RNA was known to be the carrier of instructions from the genes to the cytoplasm of the cell, one could expect to find that the RNA would be synthesized on the genes, which were located in the cell nucleus, and then transported to the cytoplasm. And indeed the few autoradiographic studies that had been done showed that the radioactive label did appear first in the cell nucleus, but then disappeared from this site and appeared in the cytoplasm. These observations were naturally interpreted as evidence of the transfer of RNA from nucleus to cytoplasm. And everyone was happy.

Now when I came to do this same experiment on macrophages, I obtained a very surprising result. I found that although the radioactive label disappeared from the cell nucleus there was no corresponding appearance of radioactivity in the cytoplasm (Plate 9). I did the experiment many times and took great care to avoid as far as possible the many artefacts that may arise when autoradiographic techniques are applied to single cells. Moreover, the measurements I made by autoradiography agreed with those made by standard biochemical techniques, so I was convinced that the observations were correct. What they showed was that a large part of the RNA made in the cell nucleus was rapidly broken down. When I did the same experiments on my multiplying fibroblasts, I saw what others had observed. The disappearance of the radioactive RNA from the cell nucleus was indeed accompanied by the appearance of radioactive RNA in the cytoplasm, but when I measured the two processes carefully I found that they didn't correspond. Neither in its timing nor in its extent did the appearance of radioactive RNA in the cytoplasm reflect the disappearance of radioactive RNA from the cell nucleus. The measurements indicated that, in the multiplying cell also, much of the RNA made in the cell nucleus was broken down. This, of course, meant that one couldn't assume, as had been done, that the redistribution of radioactivity reflected simply the movement of nuclear RNA to the cytoplasm. I wrote all this up and submitted it to the *Biochemical Journal* where John Watts and I had published our previous work on RNA turnover.

A few weeks later I received the manuscript back with dismissive reports from two referees. Both were convinced that I didn't know what I was doing. However, the Chairman of the Editorial Board, A. G. (Sandy) Ogston, who then worked in the Biochemistry Department in Oxford, didn't reject the paper out of hand. Instead he asked me to come and talk it over with him. When, in answer

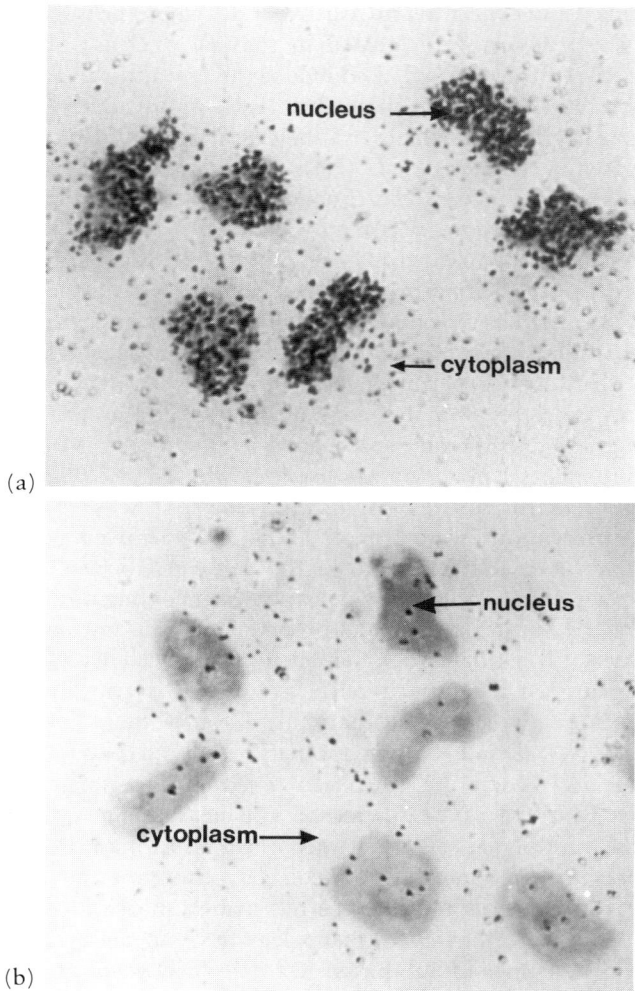

Plate 9. (a) RNA in macrophages. The cells in the upper picture show the presence of radioactively labelled RNA in the cell nuclei, revealed by the dense pattern of black grains in the overlying film. There is very little radioactive RNA (few overlying grains) in the cytoplasm. (b) The lower picture shows that, 8 hours later, most of the radioactive RNA has disappeared from the cell nuclei, but it has not appeared in the cytoplasm (no increase in the number of overlying grains). This was the first demonstration of the breakdown of nuclear RNA.

to some probing questions, I explained to him precisely what I had done and why the interpretations I had given to the results were justified, he decided, despite his referees, to publish the paper. It appeared in March 1959 and was met by a response that differed little from that of the two referees: disbelief. It seemed that there was no place in the cell, as it was then generally envisaged, for an RNA that behaved as I had described.

My natural reaction to this was to dig in my heels. I had done the experiments more carefully than anyone else, and I could see no reason why I should give way until some fault had been found in them. But the experiments were never subjected to serious criticism; they were simply rejected out of hand, essentially because they could not be accommodated within the framework of ideas that were then current. I had been relegated to the lunatic fringe and it was a very dangerous place to be. There was really only one course open to me: I had to accumulate further evidence and I had to ram it home. And thus it was that I found myself in a battle that was to consume the next five years of my scientific life.

After the collapse of Florey's plans for the Australian National University, I began to wonder what sort of future lay in store for me in England. I could see no prospect of an academic job in Australia, and although I had in many respects immensely enjoyed the years I had spent in Oxford, I didn't see much of a prospect for me there either. My salary continued to come in the form of an annual grant. There was no likelihood that it would fail to be renewed, but it hardly achieved the modicum of security that a man in his thirties with three children might have considered reasonable. There were no openings on the staff of the Dunn School and no likelihood of any arising in the foreseeable future. As far as the University of Oxford was concerned, I didn't exist, and after I had taken my D.Phil. I had no access at all to the Oxford colleges. So I gradually came to the view that I was on the market and that I should look seriously at any plausible job that came up.

My first offer of a job in England came from the Imperial Cancer Research Fund. Guy Marrian had recently been appointed to succeed James Craigie as the Fund's Scientific Director, and a handsome new laboratory building was nearing completion in Lincoln's Inn Fields. Marrian, after conversations with Florey, wrote to me to ask whether I would be interested in a position as his second-in-command to take charge of the Fund's existing laboratories at Mill Hill. Marrian was a chemist and he was anxious to have someone with a more biological background to back him up. Florey thought this an excellent job ('Plenty of money and no teaching'), so I accepted with enthusiasm an invitation from Marrian to visit the Mill Hill laboratories. After that several weeks went by before I heard from Marrian again. Then he sent a note to ask me to come to London to speak to a group of the Fund's scientific advisers. When I arrived I learnt that this group of advisers was actually an establishment committee that had been set up to make the appointment, and I found to my astonishment that I was not the only candidate. In the waiting room sat my namesake, R. J. C. Harris, whom I met there for the first time. I was asked by the Committee for my views on cancer research and replied that what cancer research needed most was a deeper understanding of the normal cell. It cannot have impressed them much, for R. J. C. Harris was appointed.

Florey was taken aback, and I was incensed, for, as I saw it, I had not applied for the job but had been offered it. I wrote a courteous letter to Marrian but one that did not disguise my surprise at the way the matter had been handled. He replied that initially I had not been in competition with anyone else, but strong pressure had developed within the Fund to expand research in tumour virology and in the end he had been obliged to bow to it. This tale has an amusing postscript. The formal letter I received from Kennedy Cassels, the Secretary of the Fund, informing me that I had not been appointed stated that a cheque was enclosed to cover travelling expenses. But the envelope contained no such cheque. I was much too annoyed about the whole performance to do anything about the missing cheque, so the Imperial Cancer Research Fund still owes me five pounds.

I was very flattered one day to receive a telephone call from Hans Krebs asking me whether I could accommodate Dr Philip Cohen of Madison, Wisconsin as a sabbatical visitor. Philip Cohen was a very influential figure in American biochemistry and many years my senior. He had now become interested in inflammation and, it appears, had been impressed by my papers on chemotaxis. He asked Krebs, with whom he had once worked, to make the enquiry. Of course I was delighted to have made such a catch, but when Philip Cohen arrived he was rather disappointed to find that I was no longer working on chemotaxis but was deeply involved in RNA. None the less he pitched in, and we had some delightful months together. Florey, disturbed about the scepticism that surrounded my recent experiments and insecure in his own judgement on biochemical matters, took the opportunity to question Philip Cohen about the work. Philip later told me that he had reassured Florey that I did in fact know what I was about.

It was in converstaion with Philip that the idea of moving to the United States was first planted in my mind. He more than once expressed surprise at the modest scale on which I operated and painted a glowing picture of what was available in America. And it was not an exaggerated picture, for the 1950s were a period of great expansion in American science and the opportunities were indeed there for the taking. I was not just then very optimistic about my prospects. The collapse of the plans for Canberra and the fiasco at the Imperial Cancer Research Fund had left me rather flat, and I was finding it very difficult to interest anyone in the observations I had made on the RNA in the cell nucleus. When the time came for Philip to return to Madison, I told him that I was open to offers.

He must have got busy pretty quickly, because within a few weeks of his departure I received two letters, one from G. Burroughs (Bo) Mider, the Director of the National Cancer Institute at Bethesda, and the other from Van R. Potter, then Assistant Director of the McArdle Memorial Laboratory for Cancer Research at Madison. Mider raised the possibility of a position as visiting scientist in the National Cancer Institute with a view to a permanent appointment in due course if all went well. Potter offered much the same thing at McArdle, but, for reasons that remain unknown to me, the offer was withdrawn a month or so later. Potter's final letter reached me on the last day of 1958.

That winter was enlivened by the presence in Oxford of George Beadle, then Chairman of the Division of Biology at the California Institute of Technology. As the Eastman Visiting Professor, he gave a series of dazzling lectures on the new molecular genetics that flowed from the elucidation of the structure of DNA. I saw a good deal of him in South Parks Road and took the opportunity to badger him with my experiments on RNA. He could offer no plausible biological explanation for what I had found, but he did accept the data. He was one of the very few who did. We had George and his wife Muriel to dinner one evening and were given a large dose of Muriel's reaction to the callously masculine society that Oxford then was. It was a preview of what later appeared in book form as *These Ruins are Inhabited*. Like Philip Cohen, George thought that I should make my way to the United States.

Negotiations with Bo Mider continued, and I waited for the administrative machinery of the National Institutes of Health to produce a formal proposal. Before it arrived, however, I received another offer of a job. This came from the Cyanamid company, which was then proposing to set up a Cyanamid European Research Institute in Geneva and was looking for someone to develop the biological side of the work. I was approached by Dr E. W. Cook, one of the company's representatives in London, and invited to visit Geneva to see for myself what was afoot. The possibility of a career in industry had never before crossed my mind but, to my surprise, Florey encouraged me to take the offer seriously. He was in a pretty gloomy frame of mind at the time and I suspect that he didn't really know what to do with me. In any case, I went to Geneva and had three very enjoyable days there, but they merely served to convince me that the Cyanamid job was not for me. So when the formal offer of a position as visiting scientist finally came from the National Cancer Institute, I accepted it. It seemed to me then that I had taken the first step towards what would eventually become the resettlement of the Harris family in the United States. It was agreed that I should arrive in Bethesda at some time in the autumn of 1959.

The fates intervened once again, this time in the form of Hans Krebs and Peter Medawar. I had had a good deal to do with Hans Krebs by this time and he was familiar with much of my work. But, as far as I can remember, I had only once met Peter Medawar. This was when he came to the Dunn School to do a collaborative experiment with Jim Gowans, and I, being in the same laboratory, exchanged a few words with him. He may well have known more about me than I thought, for he had done his D.Phil. work in the Dunn School and was on easy, although not, of course, first-name, terms with Florey. Krebs and Medawar were both at that time members of the Agricultural Research Council which distributed public funds to a number of biological and agricultural research institutes in various parts of the United Kingdom.

One of these was the John Innes Horticultural Institution then located at Bayfordbury, a small village near Hertford, a little to the north of London. The Institution had originally been endowed by

John Innes, a wealthy London business man, primarily for the purpose of training gardeners, but its first Director, William Bateson, had rapidly transformed it into a serious centre for biological and especially genetical research. It was Bateson who introduced Gregor Mendel's work to British biologists, and the word 'genetics' was indeed coined by him. In England the first plant-breeding experiments on Mendelian lines were performed at the John Innes, and the tradition of genetical research initiated there by Bateson was maintained by a succession of talented men including J. B. S. Haldane and C. D. Darlington. For many years the John Innes was the only place in England where genetics was conducted or taught, and the more enlightened British universities used to send some of their biology students there for short summer courses in this new-fangled science.

Although it masqueraded as a horticultural institution, the John Innes was a very distinguished scientific establishment. Its Director in 1959 was Kenneth Dodds and, for reasons that remain obscure to me, he was eager to develop the subject of cell biology. As far as I am aware, this was not then a recognized subject in any university in the world, and he must either have been very far-sighted or he was disenchanted with the further prospects of formal Mendelian genetics on higher plants. In any case, with the support of the Agricultural Research Council, he had constructed at Bayfordbury a new laboratory that he proposed to devote to cell biology, and it was as a possible head of this that my name had been suggested to him by Krebs and Medawar.

A letter from Dodds reached me shortly after I had accepted the offer from the National Cancer Institute. It began with a preamble about the misleading nature of the name 'Horticultural Institution' and then stressed the strong tradition that the institute had in genetics and cytology. It ended with an invitation to visit Bayfordbury in order that I might see for myself what the new cell biology laboratory had to offer. Florey was not wholly discouraging, but he clearly didn't have such faith in the unity of biology as to overcome his reservations about my going to a horticultural institution. However, he thought that I should in any case accept Dodds's invitation to visit the place.

In the course of my life I have visited many laboratories in beautiful locations, but none of them surpasses the Department of Cell Biology at Bayfordbury as it then was. The Bayfordbury estate consisted of about four hundred acres of park and woodland surrounding a graceful eighteenth-century mansion. The house, built at the top of a gentle slope, was set off by a stand of enormous Lebanese cedars, and the gardens around it were clearly the object of constant and expert care. The estate had remained in the hands of the same family until, rather neglected, it was acquired by the John Innes Trustees shortly after the Second World War. Large sums had been spent on restoring the site and on transforming the mansion into a research institute which now contained laboratories, offices, and a splendid library.

When we had had a talk about his plans for cell biology, Dodds took me on a short tour of the estate. It was a lovely late-May day and the ordered rural landscape was breath-taking. The new Department of Cell Biology stood alone beside an artificial lake flanked by large ornamental trees (Plate 10). It was a simple, elegant, two-storey building, almost ready for occupation and, to me, irresistible. All the important decisions in my life have been made very quickly, determined much more by the heart than the head. What takes the time is not making the decision, but finding the rational arguments to justify a decision that, deep down, has already been made. As Dodds showed me over the empty rooms of the new laboratory, my mind was already whirring with plans for what I would do in them. The tour ended with a walk through the national rose collection, neatly planted beside an old walled garden, and a visit to a magnificent tropical glasshouse, not actually much used for research but the pride and joy of John Newell, the estate manager. I told Dodds as I left that I would need a little time to think things over. He said I could have as much time as I wanted.

A month later Jim Gowans drove Alexandra and myself over to Bayfordbury for a closer look. This time we were also shown the staff houses that the John Innes Trustees had built in a part of the estate known as Broad Green Wood—a pretty chain of modest dwellings encircling a small green. In extent, the accommodation we were offered was not very different from what we had in Rotha Field Road, but the surroundings were incomparably more beautiful. Alexandra, although hesitant, did not seek to dissuade me from taking matters further, so I wrote to Dodds setting out what I would want in the way of staff, equipment, and running expenses. This led to a meeting in London at the end of July with William Slater, then Secretary of the Agricultural Research Council, who seemed to see no difficulty in meeting my requirements. No doubt Krebs

Plate 10. The Department of Cell Biology, John Innes Institute, at Bayfordbury, taken from across the lake.

and Medawar had already helped him make up his mind. Patronage again. A formal exchange of letters followed, and at the very end of August I wrote to Dodds accepting the position. As I was already committed to a year or so at the National Cancer Institute, it was agreed that I should begin at the John Innes when I returned from the United States.

We had booked passages on the *Queen Elizabeth*, which was due to leave on 3 September, and had been busy making ready for the journey while the negotiations with the John Innes were taking place. The Cancer Campaign, in an act of much-needed generosity, had agreed to provide travel expenses for the whole family, something that the National Cancer Institute at Bethesda was not permitted to do. I had been told that life in America was impossible without a motor car and that second-hand cars were so cheap there that anybody could afford to buy one. So I signed on for a course of driving lessons and succeeded, but only just, in passing the driving test. We leased 4 Rotha Field Road to a young American couple who were coming to Oxford for a year. This proved to be a mistake, for when we returned we found that they had done a great deal of damage. The whole of one extremely tedious day was spent at the United States Embassy in London in order to obtain the precise category of visa that each of us required. But our attitude to the trip had been transformed by the successful conclusion of the negotiations with the John Innes, for we were now going to the United States as uncommitted visitors and not as potential immigrants. We were soon to learn that that made an immense difference.

6

THE USA

There is a class of travel book that I find very hard to take. The author, usually a journalist, visits some infinitely complex place like the USA or Russia or China, spends a little time there in obviously atypical circumstances, and then feels obliged to tell the rest of the world what the country he has visited is really like. I do not for a moment imagine that the few months that I spent in Bethesda provided me with any deep insights into American life, and even now, after countless subsequent visits, I hesitate to make generalizations. The European academic at an American campus is a literary commonplace; and none of my experiences was as colourful as those of Nabokov's Pnin. But there cannot be a rich literature about Australian academics in an American environment, and certainly not about Australian academics first softened up by Oxford. So perhaps there may be a little interest in what I have to say provided that it is not taken too seriously and is seen more as a reflection of myself than as a portrait of the United States.

The Atlantic crossing was swift and uneventful. Despite the vastness of the ship, the voyage was far less memorable than the long haul from Sydney to London. There were too many people, and, on the assumption that they would otherwise be bored, a sustained effort was made to organize them. So the public rooms were always taken up with competitions, noisy party games, and bingo, from which the only escape was the promenade deck where we could gaze at the moodiness of the Atlantic and think back to the equanimity of the Indian Ocean. Like millions before us, we crowded the rail to see the Manhattan skyline emerge from the haze and the Statue of Liberty go by, disappointingly small from the deck of the *Queen Elizabeth*. We had decided to spend our first night ashore in New York city and had made a reservation at a modest hotel recommended to us by the only travel agent that Oxford then possessed. As we were thus not in any particular hurry, we stayed on board until most of the passengers had cleared their luggage through customs, and the dockside bedlam had subsided a little. Even so, my first encounter with the US customs and immigration officials was not a pleasant experience. I couldn't help wondering why they seemed so angry and why there was so much shouting. I later came to realize that in New York city in the summer everyone is on the verge of anger.

We finally got ourselves into a taxi and gave the driver, who was also incensed about something or other, the name of our hotel. He had never heard of it, but as we had the address, he agreed to take us there, though with some misgivings. When we finally reached the hotel, in a very seedy part of the city, we understood why. It had a worn and dirty look, and the people hanging about the entrance hardly encouraged confidence. But we went in bravely, announced ourselves to the surly receptionist, who demanded a deposit, and were then shown up to our rooms. The dark corridors were in acute need of redecoration and on their walls we could decipher obscene, humourless graffiti. The rooms were dispiritedly shabby and looked as if they had just been vacated but not yet properly made up. We were deliberating whether we would put up with it for one night when the sound of physical violence came pounding through the flimsy partition walls from one of the adjacent rooms. That settled it. I remembered having once been told by John French that he had stayed at the Governor Clinton, and the name had stuck in my mind. I rang the Governor Clinton and was relieved to hear that, busy though they claimed to be, they could accommodate us for the night. We ordered a taxi and checked out. The surly receptionist was not at all surprised and even refunded our deposit.

At the Governor Clinton a remarkable spectacle awaited us. As we entered the foyer we were confronted by a crowd of elderly ladies dressed in long flowing robes despite the afternoon heat, their torsos festooned with sashes of various kinds. We had stumbled upon a convention of the Daughters of the American Revolution. Even for those accustomed to the ceremonial of American public life this could be a daunting experience. We went up in the lift with a selection of blue rinses who cooed over the pink cheeks of our English children and greatly admired what they took to be their disciplined behaviour, so different apparently from that of children brought up according to the precepts of John Dewey. The Governor Clinton offered no further surprises, and when we had washed the sticky grime from our hands and faces, we set out on a little expedition to gape and marvel in the streets of Manhattan. Then supper and an early night.

We did some more marvelling the following morning as we entered the vast concourse of Pennsylvania Station from which we were to take the train to Washington. The trains looked huge to us and the parlour cars incredibly lavish, but coach class, where we had our seats, was dirty and full of people who, once again, seemed to be simmering with ill temper. A train trip from New York to Washington is not the best introduction to the United States. Even unspoiled, it must have been an undramatic strip of coast, but I can think of few regions in the world that have been so ruthlessly sacked by intensive development. The train came out of New York city as if out of a cavern, and within a few minutes the surrealist nightmare of industrial New Jersey was upon us. We simply sat there and goggled. There were indeed a few patches of green relief along the way and a longer stretch as we passed through Delaware and Pennsylvania, but these fleeting reassurances were quickly expunged as the train pulled into Baltimore. The decaying houses that backed on to the railway track there were a match for anything we had seen in the slums of England. New York had been hot and sticky, but the heat that hit us as we stepped out on to the platform in Washington was different, and it at once brought back to me the heat-waves of high summer in Sydney, a searing intensity that I had almost forgotten.

We were met by Julius (Jay) White in whose laboratory it had been arranged that I should work. He explained that the National Institutes of Health had no accommodation for visitors, even on the most temporary basis, and that he proposed to take us to a nearby motel where we could stay until we had found some satisfactory furnished accommodation. It was, he said, the only motel in the vicinity and was therefore rather expensive, but it had the advantage of being within walking distance of the laboratory. So the five of us settled into the motel and began life in the USA.

As at Oxford, my first piece of research was to find somewhere to live. Although I had been assured that this presented no problem in the United States, that certainly wasn't my experience in Bethesda. Jay White took us to see a couple of half-derelict houses that he had looked at and was rather surprised when we turned them down flat. Apparently this was the sort of accommodation that immigrants to the United States might expect to have in the first instance. We didn't see ourselves as immigrants. It was, however, urgent that we find somewhere to live. The motel, although new and efficient, was an unpleasant place. The avarice of the owner showed through not only in the room charges but also in the indifferent meals that our carless immobility forced us to take there. The owner didn't much like Europe or Europeans either. When his curiosity had prised out of us that we were not immigrants but intended to return to Europe, he offered us his opinion that Europe was a sink of decadence where the hotels were no good, the entertainment dreary, and the food indifferent. He was an adherent of the view that Europe was a place that people came from, not one that they went to. If we couldn't at once see the immeasurable superiority of the American way of life, then we must be part of the European decadence that he deplored.

Jay White was expecting me in the laboratory on the morning following our arrival in Bethesda and had made an appointment for me to see Bo Mider. The sun was already high and hot as I walked across from the motel and for the first time penetrated the immensity of the National Institutes of Health. It was then a cluster of about ten smaller buildings dominated by the gigantic edifice of the Clinical Center. The smaller buildings, each of which would have constituted a large institute by European standards, had a sedate, almost Victorian air. The Clinical Center, more recently constructed, was severely functional. I counted thirteen symmetrical rows of plain windows, but noticed that in one of the wings there were a couple of floors that didn't have any windows at all, and I wondered what they might be for (Plate 11). I soon found out.

Originally designed as a safe wing for the use of radioactive materials, this black cave now housed Jay White's department, and it was there that a laboratory had been allocated to me. The only source of air was artificial ventilation and the only source of light an array of gently flickering fluorescent tubes. I could hardly believe that people would agree to spend their daylight hours under such conditions unless it was actually necessary, but the argument seemed to be that if miners could do it then so could scientists. And indeed the scientists there had apparently become quite accustomed to their troglodytic existence and were no longer disturbed by it. I paid my courtesy call on Bo Mider, and was then introduced to my fellow cave-dwellers: Marco Rabinowitz, Joe Shack, Scotty Pratt, and Jim Reid. A little later I met two other members of the department who had their laboratories elsewhere: Mort Elkind and Si Wollman. These were to be my colleagues for the next few months.

Plate 11. The National Institutes of Health at Bethesda, Maryland. An arrow shows the windowless laboratories where I worked.

It was Scotty Pratt who by chance mentioned the Rosemary Hills apartments to me. This was quite a large development, with a rapid turnover of transient residents, and he was confident that I would be able to find a furnished apartment there. We were driven over to have a look and found a modern complex of rectangular red-brick buildings arranged like dominoes on a sloping site beside the East–West Highway. We noticed as we entered that there was a small children's playground with a swing in it. This was important because there was a swing in the garden of 4 Rotha Field Road, and Ann set great store by it. We were lucky. A two-bedroomed flat was due to be vacated in two or three weeks' time. It was equipped with simple bright ephemeral furniture and the rent was supportable. We agreed there and then to take it.

However, there was still the prospect of two or three weeks in the motel. Jay White, who seemed to dislike the motel-keeper as much as we did, had been looking into alternative possibilities. He had discovered that in down-town Washington serviced apartments could be rented on a weekly basis at a rate substantially lower than what we were paying at the motel. So we were taken to a fashionable part of Massachusetts Avenue and shown one. It was available from the beginning of the following week. The apartment was not as fashionable as the location and a little dirty, but we jumped at it. Apart from being less ruinous than the motel, it offered much-needed cooking facilities; and there was also the agreeable prospect of a spell of sightseeing in Washington for Alexandra and the children. So, a couple of days later, we became residents of what we were told was known locally as Embassy Row.

Our stay there was pleasant and uneventful. There was a small bus that ran along Massachusetts Avenue to the very front door of the Clinical Center. I took it to and from work each day and thus learned that in Washington public transport was used only by those at the bottom of the social scale, mainly black people. Their conversation was always about simple, practical things, the preoccupations of those for whom life was not easy. A small supermarket close to our apartment initiated us into the technique of keeping house in the USA. Supermarkets had not yet made their appearance in Oxford. During the week Alexandra and the children made some tentative explorations of Washington, and I joined them at the weekends. The parts of the city that we visited were very handsome: wide tree-lined avenues, parks, splendid monuments, opulent museums. But we were told that there were some areas, predominantly black, that it was dangerous to visit. This to us was full of menace, for many years were to elapse before the same could be said of inner-city areas in England. We found the architectural contrasts and the abrupt transitions of Washington a little bewildering; and we never did find out why, in one direction, the streets were designated by number up to a certain point and then suddenly thereafter by letter. When our apartment at Rosemary Hills became vacant, we moved with a sense of relief, for although the high-life on Embassy Row was enjoyable enough, we all felt it was time that the peaceful normality of our lives was restored.

There was, however, still one obstacle to be overcome. A small bus did run at regular intervals from the apartments to the shopping centre at Silver Spring, but it was clear none the less that life would be very difficult without a motor car. Jay White had a very senior technician who was an expert in motor-car transactions, and it appeared that he had just the car for me. It was an aged Chevrolet, but, of course, in impeccable condition and going for a song: two hundred dollars. He drove me out into one of the peripheral suburbs of Washington to see the object of his praises, and there, parked in front of a modest house, stood a dark-blue machine with a long narrow tail. I suppose that my ignorance in the matter of motor cars was at that time unrivalled, but my adviser was very persuasive and I took the plunge. He suggested that I should drive my mechanical masterpiece back to the National Institutes of Health by following his car. Easier said than done. My driving experience was limited to the lessons I had taken; I had never handled a large American car with the gear lever on the steering column; and I had never driven on the right-hand side of the road. The car I was meant to follow lost me almost at once, and I had only a very general idea of the way back to the National Institutes of Health. What followed was one of the most hair-raising experiences I have ever had, but in the end, dripping with sweat and an hour late, I made it. I now had a fixed abode and a motor car. I was ready for life in America.

I must confess that the experimental work I did at the National Cancer Institute was of no consequence. Since I was to be there for only a year at the most, there was no point in my attempting to set up elaborate new laboratory techniques or to duplicate the activities that John Watts and Marianne Jahnz were continuing to pursue at Oxford in my absence. What I hoped was that I might be exposed to other scientists and learn something of the experimental approaches that they used. Curiously enough, this was not easy to achieve at the National Institutes of Health. Although there were vast numbers of scientists there, they appeared to operate within units that were almost hermetically sealed. There was very little communication between one laboratory and another, and on several occasions I found that the inhabitants of one unit were not even aware of what was going on in a unit on the floor below or at the other end of the corridor. It seemed to me that the whole establishment was much too big for effective communication to be achieved without special effort. Moreover, working practices militated against the spontaneous free flow of information. People would arrive in a great wave between eight and nine o'clock in the morning and leave in a great wave at about five o'clock in the afternoon. Very few lights stayed on in the laboratories at night, and one had the impression that the ethos of the place was that one should do an honest day's work and then go home, just as one would in any other branch of the civil service. There were occasional rushed lunchtime seminars, arranged informally for small groups with a common interest, but on the whole people didn't talk too much about their work. The atmosphere was totally unlike the creative babble that I was used to in South Parks Road. It was much more like working in a factory than in a university.

Among the cave-dwellers there were two whose work was of special interest to me. Joe Shack was one of the earliest investigators of the physico–chemical properties of DNA, and he had a number of very interesting ideas about how this extraordinary molecule might be replicated and transcribed into RNA. Marco Rabinowitz was working on the synthesis of proteins in reticulocytes. Reticulocyte is a name given to a certain stage in the development of the red cells of the blood. The precursors that give rise to red blood cells originally have nuclei, but as the cells mature, the nuclei are eliminated and one is left with cells composed of cytoplasm only. These none the less continue to develop further and to synthesize protein, mainly haemoglobin, the compound that carries the oxygen in the blood. The anucleate cells that synthesize haemoglobin are called reticulocytes because when they are stained in a particular way they show an internal network, or reticulum.

I have already mentioned the intense interest I had in the work of Hämmerling and the observations he had made on *Acetabularia* cells from which he had removed the nuclei. Now here in the reticulocyte we had another cell-type that had lost its nucleus, but one very different from the giant single-celled seaweed. It was therefore of great interest to me to compare what was going on in the two cell-types and to see what they had in common and in what ways they differed. The interpretations that I later gave to some of my own experiments on RNA were profoundly influenced by what emerged from the study of these anucleate cells. I spent a good deal of time with Marco and Joe and we usually had lunch together at a small nearby restaurant which, although less convenient than the canteen at the National Institutes of Health, offered a line in roast-beef sandwiches that met with Marco's approval.

Mort Elkind, although a member of Jay White's department, was not a cave-dweller but worked in a laboratory with windows a couple of floors below. Some time elapsed before I got to meet him and found out that he was working on a problem that was of importance to any cell biologist with an interest in cancer: the biological effects of radiation. In fact he was making such precise measurements of the lethal effects of radiation on mammalian cells that his experiments eventually became classics of radiation biology. We have continued to meet from time to time, and it has given me pleasure to see this sincere and unaffected man rise to a position of great distinction. There were people then working at Bethesda whom I would very much have liked to meet, had I known they were there. Marshall Nirenberg must at that time have been laying the foundations of the work that was soon to enable him to decipher the genetic code, but I didn't actually meet him until many years later when I visited the National Institutes of Health to give a lecture. Leon Heppel was also there. Almost everyone who worked on the biochemistry of DNA or RNA in those days owed something to Leon Heppel, for the compounds that he synthesized were essential tools for a very wide range of experiments in that area. I regret that I never did get to meet Leon Heppel. But, by chance, I did meet Harry Eagle and Gordon Tomkins.

Harry Eagle was heavily involved in what was to become the definitive study of the nutritional requirements of mammalian cells in artificial culture. Harry knew about my work on the incorporation of radioactive amino acids into cellular proteins and was especially interested in my observation that fibroblasts did not survive unless certain forms of sulphur were added to the medium. He later explored this phenomenon further and published a couple of interesting papers about it. In his laboratory there was a young man named James E. Darnell, Jr. who was studying the effects of the poliomyelitis virus on the synthetic machinery of the cell. I found his work interesting and induced him to come and give a lunchtime talk on it to the denizens of the cave. A few years later, Jim Darnell and I found ourselves on opposite sides of the fence in the controversy about RNA turnover, but the disagreement didn't last too long.

Gordon Tomkins was a most amusing and talented individual. He was then initiating the research for which he later became well-known. This centred on the mechanisms by which the cell regulates the synthesis of enzymes, and since this was a subject in which I also had a stake, we had a lot in common to talk about. Conversation with Gordon always had an element of vaudeville about it; and his premature death deprived us all of a rich source of light-hearted generosity, a commodity that has become very scarce in contemporary science.

People were very hospitable, and most of those whom I have mentioned invited us to their homes at one time or another during our stay. The disparity between English and American academic salaries was not then as great as it is now, and I did not see in the homes that I visited the glamorous affluence that Hollywood portrayed as the American norm. The laboratory chiefs certainly lived in large comfortable houses, but not notably grander than the homes of some of the professors in Oxford. Most families had two large cars, but all American cars were large, and in the Bethesda region public transport was so rudimentary that cars were as indispensable as toothbrushes. Scientists who had not yet reached senior executive positions lived modestly, although no scientific position at the National Institutes of Health entailed the conditions of near poverty that were imposed by the lowest rungs of the academic ladder in England. Of course, you couldn't spend a day in the United States without seeing flamboyant exhibitions of great affluence, but I didn't come across any examples of this among the scientists whom I met.

There was one respect in which their lives seemed much less secure than those of English academics. The National Health Service in England, though of course imperfect, did function reasonably well and no one worried about the financial consequences of ill-health. This seemed, however, to be a constant preoccupation of my colleagues at the National Institutes of Health. The various private insurance schemes were said to be expensive and didn't apparently offer adequate cover for protracted illnesses or those requiring sophisticated hospital procedures. The financial threat of prolonged illness seemed very real. We only had one occasion to use the services of a doctor while we were in America; his bill came promptly and it was indeed, by our standards, exorbitant.

There also seemed to be a great deal of concern about the rising costs of higher education. Salaries at the National Institutes of Health were on civil service scales and the scientists who worked there were not free to supplement their income in ways that were open to those employed by universities. Apparently civil service salaries in that range did not easily encompass the university education of three or four children. In England where entry to the universities was competitive, the worry was that your children might not get in, but if they did, the system of grants introduced by the government after the War took care of the finance. Our overall impression was that life in England, despite the financial limitations, was less risky.

Living in the Rosemary Hills apartments soon brought it home to us that Maryland, although its original character was much diluted, was still part of the South. When I left Australia there was no colour problem there. The aborigines had been all but wiped out by the settlers except in some remote and, for most Australians, inaccessible regions in the interior. In Sydney, apart from a few families who sold boomerangs to tourists at La Perouse, aborigines were rarely to be seen; and a strictly applied 'White Australia' policy ensured that permanent immigrants from 'non-white' countries were a rarity. In Britain a sizeable movement of people from India, Pakistan, Africa, and the Caribbean had taken place since the War, but the immigrants had settled almost entirely in London and the industrial cities. The only black faces you saw in Oxford were those of the few privileged individuals who had come to study there. Bethesda was our first contact with a black population, if you could call it contact. In Maryland the white residential areas were pretty sharply segregated from the black. Each morning large numbers of black people moved into the white areas to do, as far as I could see, mainly menial tasks, and each evening they returned to their own enclaves. There was no significant traffic the other way. Indeed, on the few occasions when we happened to stray into black areas either on foot or in our long-tailed Chevrolet we could sense the hostility.

However, I certainly didn't feel this in the very small number of black people I met personally. There was the cleaner in the laboratory who initially regarded me with great curiosity but soon formed the habit of stopping by each day for a chat. The man who delivered the post to our rooms did much the same, although less regularly. It seemed as if the fact that I was not American made relations easier rather than the reverse. The lady who cleaned our flat while we were on Massachusetts Avenue went out of her way to be helpful; and at the Giant Food supermarket in Silver Spring where we stocked up each week, the young men who piled what we had bought into the boot of our car always seemed to have a friendly remark for us. I told Joe Shack and Marco Rabinowitz at lunch one day, only half in jest, that the nicest people in Bethesda were black. They didn't disagree.

Rosemary Hills School, where Paul was enrolled a day or two after we had moved in, was virtually all white even though it was located within walking distance of a black area. A few black children had been admitted (I was told this was called token integration), and there was one, Jerome, in Paul's class. Paul and Jerome became firm friends, drawn together by the fact that they were both objects of intermittent persecution, in Paul's case mainly because he had, by the crew-cut standards then obligatory, long hair and spoke with what in the USA is known as an English accent. In Oxford Paul had begun at a gentle little elementary school called Greycotes and at first he was bewildered by the sudden transition to the rather heavy-handed style of Rosemary Hills. However, the head master did tell me that Paul derived great pleasure from the freedom to lace his lunch with as much tomato ketchup as he pleased.

After he had been there for a while, we asked Paul to invite Jerome home for afternoon tea. We greatly enjoyed having him and he did not seem in the least ill-at-ease with us. Another black boy called Dexter, who was a little older, came to fetch Jerome and, before they left, the two of them took Helen and Ann down to the playground to give them a push on the swings. We asked them to come again, but they never did. We were very slow to realize that the Rosemary Hills apartments were 'segregated' and that the sight of Jerome and Dexter in the playground with our young daughters had not pleased our neighbours. What steps were taken to ensure that Jerome and Dexter didn't come again, I don't know.

There were some aspects of life in Bethesda that reminded me of Australia. The pressure to conform was very strong. Our neighbours in the Rosemary Hills apartments were intensely curious about us, and at each new contact we were subjected to an uninhibited battery of personal questions aimed at establishing precisely who and what we were. Until this was clear, relations remained uneasy. People were also very eager that we should join in the various forms of organized social activity that seemed to take up much of their time; and an invitation declined could easily give offence. It was not simply a matter of hospitality, but an intuitive dislike of strangers who stood off and did not want to be immediately assimilated. Everyone displayed a recognizable religious affiliation, and I had the impression that, at least in public, disbelief was equated with subversion. One of our neighbours asked me point blank one day what my religious beliefs were. When I told her light heartedly that I was an atheist, she was obviously appalled and thereafter would have nothing to do with us.

These attempts to impose conformity, usually attributed to the 'melting-pot' character of American life, were something I was already very familiar with. In Australia there was a similar preoccupation with the defence of the national identity even though the extent of non-British immigration before the War was trivial. What struck me when I came to Oxford was that nobody cared at all whether one conformed or not. To ask personal questions on scant acquaintance was simply tactless, and questions about religious beliefs were never asked at all. Eccentricity of dress or behaviour was not only tolerated, it was hardly noticed. Indeed, in the eyes of young Americans or Australians in Oxford, eccentricity seemed to be the rule. Newcomers were not pressed to join anything or to adopt English habits or to modify their style of life in any other way. So completely was privacy respected that visitors from overseas often concluded that the English were inhospitable; and it was to meet criticisms of this kind that a Newcomers' Club was later formed in Oxford essentially to offer visiting Americans some semblance of the reception that they themselves offered and expected. The difference between arriving in Oxford and arriving in Bethesda was profound. It was as if in the new world national identity had constantly to be reasserted, whereas in the old it was taken for granted.

An Australian visitor to the USA in 1960 would have felt quite at home with many features of American life: the informality, the immediate use of first names, the spontaneity and simplicity of social relationships. He would have shared an admiration for things new and modern. Unlike an Englishman, he would then have preferred a new house to an old one, and he took pleasure in dramatic and ostentatious architecture. Practical considerations loomed large in his mind: an up-to-date kitchen and bathroom and a large swimming-pool impressed him more than weathered stone or old beams. He was not usually so committed to the pursuit of wealth as his American counterpart, but he gave affluence due respect. He was keenly interested in new gadgets, new motor cars, new and more efficient ways of doing things; and he was impressed by size. A European in America might have missed the deep reassurance of ancient towns and ancient traditions; but an Australian would have been perfectly at ease with impermanence, high mobility, and a rapid rate of change.

I once went up in an elevator with a young American scientist to whom I had been introduced a few days previously. I asked him how he was. 'Fine', he said, 'just fine', and struck a pose that indicated robust health and great energy. And how were his experiments going? 'Fantastic', he answered and launched into a string of superlatives about the observations he had just made. A young Australian would have responded in much the same way. A young Englishman, however, feeling equally vigorous and equally enthusiastic about his work, would have assured me that he could hardly walk up the stairs and had done nothing but break glassware for months. Had I come to the USA directly from Australia I think my reactions would have been much the same as those of the hypothetical Australian I have been describing. But the seven years in Oxford, some of them pretty lean, had clouded my antipodean vision and I viewed Bethesda though spectacles that were already half-English.

Autumn in Oxford is very beautiful, but it is a wistful, almost resigned beauty shrouded in mists and wet with intermittent rain. The Cotswolds turn various shades of brown with only hints of pink and yellow. The fall on the north-eastern seaboard of the United States has a quite different character. The air is clear and bright, and the trees make great dramatic splashes of red and yellow. It is an exhilarating time, not drenched in nostalgia as it is, except for an occasional day, in the Thames Valley. Just as, for me, Europe offers nothing finer than alpine meadows in spring, so America offers nothing finer than the fall in New England or Vermont. Maryland is not Vermont, but in a more subdued way it is still a place where the fall can exercise its magic; and the coming of the cooler weather found us for the first time in possession of our own independent means of transport.

The long-tailed Chevrolet gave us a freedom that we had never previously possessed. First, little trips in and around Washington: more of the museums, Rock Creek Park, the Potomac. Then, as my confidence in the Chevrolet grew, longer trips: to an immediately familiar eighteenth-century Annapolis; to Fredericksburg and the battlefields of the Civil War; across the seemingly endless bridges to the other side of the Chesapeake Bay. One thing that struck us in all the museums, all the historical sites, even the parks, was the care that had been taken to display simple explanatory material for visitors, and especially for children. It was as if every notable place to which the public had access was part of the educational system. We found this admirable. On one occasion we drove a little further into Virginia to visit Jamestown and the restorations of Williamsburg. This was a very confusing experience. On the one hand, infinite care had been taken to preserve and display the roots of American democracy; on the other, the doors of restaurants and public lavatories displayed notices indicating that they were for the use of whites only. There are some things that are difficult to come to terms with. We did not venture any deeper into the South.

I have said so little about the work I was doing in the laboratory that the reader might be tempted to conclude that I wasn't doing any. Well, as I have already indicated, I didn't do anything worth writing up, but I wasn't exactly bone idle. My mind remained fixed on the problem of RNA turnover and its possible biological significance. I was now completely convinced, whatever anyone else might say, that much of the RNA synthesized in the cell nucleus was rapidly broken down; but I didn't know where it was broken down. Why was this important? For the following reason. If this short-lived RNA was transferred from the nucleus to the cytoplasm before being broken down, then it had to be involved in some cytoplasmic activity that required its rapid consumption. In all probability this would be some activity associated with the synthesis of protein, as the only functions then known for RNA were associated with protein synthesis in some way. If, however, the short-lived RNA was broken down within the nucleus, where it was probable that no synthesis of protein occurred, then some completely new (and totally obscure) function for this RNA would have to be envisaged.

This was the question I fiddled with for most of the time I spent in Bethesda. I wasn't able to devise a decisive experiment, but I did accumulate some wisps of information that moved me closer to the view that the breakdown of the RNA took place within the cell nucleus. This was an even more dangerous position than the one I was already in. Waiting for the elevator one day (there was a good deal of waiting for elevators in the Clinical Center), I was joined by Bo Mider who had time to inform me that my experiments on RNA were not accepted by some of his colleagues. I thanked him for the information and wondered what his reaction might be to the even more eccentric position that I was beginning to adopt. If people found it difficult to accept that much of the RNA made by the cell was rapidly broken down, what would they make of the idea that it was broken down almost as soon as it was made and for no apparent purpose? If it turned out that further and better experiments supported this idea, I was obviously in for a hard time.

Not being completely obsessed by my experimental work, I took the opportunity offered by the marvellous library facilities at the National Institutes of Health to do some heavy reading in two areas about which I knew very little. The first of these was botany. The impulse here was not only the fact that at the John Innes I would be surrounded by botanists. I had the feeling that the extraordinary variety of plant forms might offer material that would lend itself more readily to the investigation of certain kinds of problems than the comparatively restricted range of forms presented by animal cells. *Acetabularia* was a case in point. This reading did not of course make me a competent botanist, but by the time I returned to England I felt at home with botanical language and had a pretty good idea of what the plant world had to offer.

The other subject in which I immersed myself was genetics. My motivation in this case derived from my continued interest in the cancer problem. Whatever it was that distinguished a cancer cell from a normal one, it was a property that the cancer cell passed on to at least some of its descendants: cancer cells beget cancer cells. If they did not, we would not be worrying about them. It was obvious that at the level of the cell we were dealing with an aberration that was transmitted from one cell generation to the next. Now genetics was the science that dealt with the analysis of traits transmitted from one generation to the next; and although nobody in 1959 could see how genetical methods, which required at the very least that the organisms being studied should engage in some kind of mating process, could be applied to the cells of the body, these methods had reached a high level of sophistication and I thought I had better take them on board. That proved before long to be a very fortunate decision.

Apart from science I read mainly American history; but my year was made by the appearance of Richard Ellmann's biography of James Joyce. There were no good bookshops in Bethesda or Silver Spring, but by chance I saw this book in a shop window and bought it at once. Joyce's *Ulysses* had been a bed-book of mine in undergraduate days, and Ellmann's account of his life was a landmark. Some years later Ellmann was elected into the Goldsmith's Professorship of English Literature at Oxford and I was proud to have him as a colleague.

We had a visit from the Mackaness family while we were in Bethesda. George was still on the staff of the John Curtin School in Canberra and was spending a sabbatical year at the Rockefeller Institute in the laboratory of René Dubos, an eminent bacteriologist (and also, incidentally, a very entertaining writer). It appeared that George was not having an easy time of it with Hugh Ennor as Dean, and he soon left Canberra for a position in Adelaide. That wasn't a great success either, and after a couple of years there he accepted the Directorship of the Trudeau Institute for Medical Research in upper New York State. This move to the USA turned out to be irreversible, and George, like myself, became a permanent expatriate.

A couple of months after his visit to Bethesda, I spent a day with George at the Rockefeller Institute where he was busy pursuing his investigations on the part that macrophages play in the defences of the body against infection. I think he was responsible for introducing into the Rockefeller Institute certain approaches to the study of these cells that continue to form part of experimental programmes there to the present day. I managed to have a few words with René Dubos, who was interested in the subject of chemotaxis, and was introduced to Peyton Rous whom I have already mentioned. The Presidential Election of 1960 was not far off and the American public was then being entertained by the spectacle of the Kennedy–Nixon debates. (My reaction to these performances was dismay at the thought that one or other of these two candidates was going to be the President of the United States.) The Rockefeller Institute was a stronghold of the Republican Party, but Peyton Rous sported on his lapel a huge rosette advertising his allegiance to the Democrats. I went home with George at the end of the day and found that he lived in a small dingy apartment at the top of several flights of narrow stairs. Even as a temporary arrangement it must have been depressing after the semi-rural ease of Canberra.

Alex Leaf organized my first visit to Harvard, and I went all the way up to Boston by train mainly so that I could look out of the window. I felt immediately at home in the Harvard environment: the jumble of old and new buildings, the intellectual ferment, the preoccupations peculiar to academics. Alex had arranged for me to spend most of my time with Paul Zamecnik and his colleagues, which turned out to be not only a rewarding experience from the scientific point of view but also a great personal pleasure. Paul Zamecnik was, and still is, a remarkably gifted scientist whose contribution to our understanding of the way in which proteins are synthesized cannot be overestimated. He had discovered what is known as 'transfer'

RNA, a family of RNA molecules whose function is to pick up the individual amino acids and align them in such a way that they can be linked together to form the amino acid chains of which proteins are constituted. In the study of protein synthesis, a more important discovery could hardly be imagined. But there was a great deal more to Paul Zamecnik than his eminence as a scientist. Eminent scientists come in all shapes and sizes, and many of them are not particularly likeable. But Paul Zamecnik was a treat: sensitive, unassuming, generous—a model for us all.

I gave a seminar at the Massachusetts General Hospital in what is known as the Ether Dome, the theatre in which ether was first used as an anaesthetic. I talked about the RNA in the cell nucleus and, for the first time, was listened to with care. John Littlefield, Bob Loftfield, and Jesse Scott were then members of Zamecnik's department, and after the seminar, the three of them, together with Zamecnik, discussed my experiments with me in great detail. What a delight! Jesse Scott said he was going to repeat my experiments with some cells that he was working with, and a few months after I had returned to England, I received a card from him telling me that he could confirm my observations and agreed with the interpretation I had given them. For me that was a monumental document: the first crack in the wall of disbelief. John Littlefield and Bob Loftfield both spent a period of sabbatical leave at the Dunn School after I had been elected to the Professorship there. My stay in Boston was much too short, and I left it not only with regret but with the thought that if I had come to Harvard instead of Bethesda, my impressions of America and perhaps even my ultimate fate might well have been very different.

Bill Barclay invited me to Chicago. I booked a sleeper on the Baltimore and Ohio overnight train from Washington and was amazed to find that the sleeping compartment was an array of narrow metal boxes into which the passengers were squeezed like the proverbial sardines. The train arrived in the very early morning, but Bill was there to meet me and had generously set aside the whole of the day for my entertainment. I earned my keep by giving a lecture in his department, but there wasn't much interest there in RNA and, duty done, the rest of my stay was given over to sightseeing. There was a curiously rancid smell in the air and Bill explained that when the wind blew from the stockyards that was the smell you got.

Most American cities show stark and closely apposed contrasts, but Chicago was quite extraordinary. The University of Chicago, a richly endowed, distinguished, and in a way rather idealistic institution was situated in an area of such squalor and violence that the staff had taken to arriving and leaving in cohorts in order to avoid being mugged. No novelty now, but very unusual in 1960. The views from the famous overhead railway included some of the worst slums you could find, but the tall buildings along the lake front presented a vista of serene and sparkling elegance. The city throbbed with the pulse of American business at its most aggressive, but the museum of fine arts contained a dazzling collection of paintings, chosen and displayed with the greatest discrimination. My visit ended with dinner at the Stockyards Inn, originally an eating house for the cattlemen who brought the beef on the hoof to Chicago for slaughter in the stockyards, but now an elegant and expensive restaurant. The waiters wore what was meant to be eighteenth-century attire. It was the first place I had been to where you ordered your prime steak by weight.

I flew into San Francisco in the evening and from my window seat saw the Sierras suddenly turn pink on one side as the sun went down. My tour of the West Coast had been arranged by George Beadle; he had apparently been able to induce Harry Rubin at Berkeley and Charles Yanofsky at Stanford to be kind to me. I knew them both only by name, and I don't suppose either of them had heard of me. But they were indeed very kind, and they listened patiently to what I had to say. Before crossing over to Berkeley I had a couple of days of unrestrained tourism in San Francisco itself. It is the favourite city of countless Americans, but after Sydney it was very small beer. San Francisco bay is not nearly as picturesque as Sydney harbour, and Chinatown was no novelty to anyone familiar with pre-war Sydney. What struck me most was the layout of the streets. In most cities built on hills the streets wind around the hillsides in order to reduce the steepness of the gradient, but in San Francisco the hills are completely ignored and a perfectly symmetrical grid has been plastered on top of them. Some of the streets are so steep that you feel as if you are climbing rather than walking up them; and a new vista emerges abruptly when you reach the top.

On my first evening I went for a stroll around the centre of the city and ended up in a small restaurant that seemed to be a haunt of taxi-drivers. As I sat eating my hamburger, a driver came in and

joined three others at a nearby table. I could overhear their conversation. The snippet that follows is engraved on my memory:
  'You heard about Joe? He's dead.'
A long pause.
  'Had cancer, didn't he?'
  'Yeah.'
  'Got any family?'
  'Wife. Two kids. Still at school.'
Another long pause.
  'She got any money?'
  'You ever hear a somebody die a cancer got any money left?'
I thought of the National Health Service in England and wished that its critics could have heard that. The following day I was taken to dinner by Henry Cohn, a distinguished radiobiologist whom I had met one evening in Bethesda at Mort Elkind's home. He drove me to a swish restaurant that looked out over the Pacific. It was there that I first became acquainted with the dimensions and complexity of a Caesar salad.

George Beadle had sent a car out to the airport in Los Angeles to drive me to the California Institute of Technology (known worldwide as Caltech) in Pasadena. There was a bit of smog in the air in Los Angeles, but it seemed to get worse as we drove out, and by the time we reached Pasadena everybody's eyes were watering slightly. George had laid on a couple of marvellous days for me. I met and had lively exchanges with several people who were, or were destined to become, famous names in biological science: Bob Sinsheimer, Ed Lewis, Ray Owen, Matt Meselson. After my seminar, the legendary Max Delbrück simply asked me why I chose to concern myself with biochemical minutiae. Renato Dulbecco, however, seemed to be genuinely interested, though his eyes were streaming and he was obviously greatly troubled by the smog. I was not surprised to learn a couple of years later that he had gone elsewhere. There were reminders of Australia in Pasadena. Australian gum trees had been planted in many parts of California and thrived there. I also saw some banksias and bottle-brushes; and orange groves, something I hadn't seen or smelled since coming to England. The garden beds around the laboratories were full of bright red cannas, just as they might have been in Sydney. But for the smog, I think I could easily have naturalized in Pasadena.

On the way back to the East Coast I stopped over in Denver to see Ted Puck, who had recently moved there from Caltech. This too had been arranged by George Beadle. Ted Puck had devised what became the standard method for cloning animal cells (growing colonies from single progenitors), a development of crucial importance for cell biology. He and his colleagues were extremely hospitable, and I was again taken to dinner in a restaurant with eighteenth-century waiters. Ted visited me soon afterwards at the John Innes, and, over the years, a very rewarding exchange of scientists has gone on between our two laboratories.

Spring came suddenly. We saw the cherries flower in Washington and, more beautiful, the white and pink bracts of the North American dogwood. But we were all getting rather restless, and the coming of the warm weather gave notice of the fierce Washington summer. There seemed to be no overriding reason why we should stay to sweat it out, and I was becoming increasingly impatient to get down to some real work. John Watts and Marianne Jahnz had already transferred to the John Innes and wrote to let me know that things were moving there. I wrote to Kenneth Dodds to have his reaction to the idea that I might return sooner than expected. The sooner the better as far as he was concerned, although it looked as if we wouldn't be able to occupy the house in Broad Green Wood until September. We found that berths were available on the *Queen Mary* for a crossing at the beginning of June and made the booking. I sold my long-tailed Chevrolet for one hundred dollars and made my peace with the tax inspector in Washington. We returned to New York by train, but this time felt affluent enough to afford ourselves the luxury of a parlour car. At one point the children started chattering about what they would do when they got home, and then at last it dawned on me that for me too, home was no longer Australia.

7

THE GREEN WORLD

Home for the next three years was Broad Green Wood, Bayfordbury, but it took us a couple of months to get into it. When we returned to England, the house on the green was not yet ready for occupation and our own home in Oxford had been leased until the end of August. So for a few months we found ourselves once again in makeshift accommodation. To begin with, Alexandra and the children made do in a boarding-house in the Banbury Road in Oxford, while I put up at a hotel in Hertford, the town nearest to Bayfordbury. The savings from my American stipend enabled me to buy a car, a new, gleaming black Morris saloon which I drove in style to Oxford for the weekends. When the house on the green became available, we camped in it until our furniture was released by the departure of the young American couple from 4 Rotha Field Road, which we then put on the market and soon sold. After what seemed an eternity of dislocation, September found us all together again with our own furniture in a freshly decorated house on the edge of a dense wood in the depths of rural Hertfordshire. *Nel mezzo del cammin di nostra vita mi ritrovai per una selva oscura.*

Each morning I drove Paul and Helen, and in due course Ann also, to a small dame school in Hertford where, as far as I could see, they were happy and their education was not actually impeded, then brought the car back to Broad Green Wood and walked to work down a woodland path that led from the green to the laboratories. At the end of the day I walked back along the same path transfigured in the evening light. When the weather smiled, we would drive to London or Cambridge or to places of interest near by. There was nothing to distract me from my work or to disturb the peaceful tenor of our daily life. Heaven.

I did not at once set about organizing the beautiful new department that had fallen to my care. I do not believe, and did not believe then, that creative scientific institutions can be organized into existence. The managerial skills needed to set up a new supermarket are not enough, perhaps even inappropriate. What makes a creative scientific institution is, of course, creative science, but also something else, something more closely akin to the devotion that eventually brings to maturity an exotic and delicate tree. Like the tree, a research institute needs to be fertilized, pruned, but with great restraint, shielded from inclement winds; and eventually, with a little luck and a great deal of judgement, a thing of great beauty emerges.

I had inherited three excellent biologists who knew a great deal about plant cells, a biochemist, and a physicist with a special interest in advanced microscopic techniques. I saw at once that the skills that these people possessed could be of help to me in the pursuit of my own research, but I did not seek to change the direction of their work. Collaboration is something that has to develop spontaneously, it cannot be imposed. Nor did I see any great urgency to fill the positions that had been made available to me. What did seem to me to be urgent was that I should get my own experiments going. It appears that I had already acquired what was to become a life-long conviction that if my experiments went well everything else would fall into place. John Watts and Marianne Jahnz had brought some home-made essentials down from Oxford and had already got the cell cultures going; John had in fact made some important improvements to the techniques we had been using. Within a few days of my arrival we were hard at it, probing what I then regarded as the most important question in the field of RNA metabolism: where in the cell was the short-lived RNA broken down?

It might be imagined that the answer to this question should not have been too difficult to obtain. Let me assure the reader that with the methods then available it was fiendishly difficult. The techniques for separating the individual RNA components in the cell (fractionating the RNA) were crude and laborious. They were all based on prolonged high-speed centrifugation, were very insensitive, and carried a high risk that the RNA might undergo degradation during the procedure. The techniques for separating the different structural elements in the cell (fractionating the cell) were even cruder. They involved breaking the cell up in a mechanical blender or homogenizer of some kind and then separating the individual components in the resulting mess, again by centrifugation. During the process of cell disruption, larger structures such as the cell nucleus were naturally seriously damaged, and there was gross contamination of the nuclear fraction with cytoplasmic components and vice versa.

The whole investigation was a constant uphill struggle against experimental artefact.

John Watts, refining some observations that we had found in the literature, developed a rapid one-step procedure for separating the RNA in the cell nucleus from the RNA in the cytoplasm. This procedure involved treating the cells with water that had been saturated with phenol: this precipitated the nuclear RNA but left the cytoplasmic RNA in solution. The separation was not perfect, but it was simple and reproducible, and it permitted us to repeat our work on the turnover of the nuclear RNA with much greater precision than we had previously been able to achieve. The results did not budge. There was no doubt that much of the nuclear RNA that was labelled by a brief exposure of the cells to a radioactive RNA precursor was rapidly broken down; but something much more ominous emerged. We couldn't find any RNA like that in the cell cytoplasm.

If the short-lived RNA was broken down within the cell nucleus, then we had obviously stumbled upon a phenomenon of great biological importance, but it was one that we would have a great deal of difficulty in selling to a sceptical world, especially as we had no idea of what the functional significance of the phenomenon might be. However, our evidence that the breakdown of the RNA took place within the nucleus was still very weak. It was altogether possible that this RNA had certain special physical properties that caused it to be precipitated with the nuclear fraction in our separation procedure, even though in the intact cell it might have been present in the cytoplasm. It was clear that we would have to devise a new technique for separating the nucleus from the cytoplasm and one that minimized or, hopefully, eliminated the possibility that, in some way, we were losing the short-lived RNA from the cytoplasmic fraction.

I have always greatly enjoyed developing new techniques and on this occasion shared the pleasure with Harold Fisher who was spending a sabbatical year with us on leave from Ted Puck's department in Denver. The technique we finally hit upon was delightfully simple. We found that if you stirred the cells under strictly controlled conditions in a very dilute solution of one particular kind of detergent (not the sort you use in the kitchen), then the membranes in the cell were gently dissolved and the nuclei popped out unbroken. The detergent prevented any cytoplasmic components from adhering to them. The nuclei could then easily be separated off and washed repeatedly in fresh detergent solution in order to collect any possible traces of contaminating cytoplasm. We explored the limitations of the method with great care and finally convinced ourselves that if the cytoplasmic fraction prepared in this way didn't show any RNA having the characteristics of the short-lived RNA that we had found in the cell nucleus, then that RNA wasn't there. It was at this point that I received the encouraging note from Jesse Scott telling me that he had been able to confirm the work that I had talked about in the Ether Dome at the Massachusetts General Hospital.

While we were doing all this, short-lived RNA became a very hot topic in the biological world. Most of the RNA in the cytoplasm of the cell is present in the form of small particles called ribosomes in which the RNA is packaged into a complex with some fifty different proteins. Before 1961 it was generally assumed, although no specific experiments had been done to test the idea, that the RNA that carried the information (the genetic specifications for the synthesis of particular proteins) from the genes to the cytoplasm of the cell was the RNA in the ribosomes. In June 1961 there appeared in the *Journal of Molecular Biology* a remarkable paper by François Jacob and Jacques Monod entitled 'Genetic regulatory mechanisms in the synthesis of proteins'. It was very different in quality from the papers from Monod that I had previously studied. The association with Jacob had obviously modified Monod's thinking profoundly. The centrepiece of the paper was a review of Jacob's marvellously imaginative genetic analysis of the mode of action of regulatory genes (genes that control the action of other genes). But superimposed on these genetic experiments there was a massive deductive argument that bore the hallmark of the Monod I was familiar with.

The principal theme of this argument was that the synthesis of proteins was regulated not in the cytoplasm of the cell where they were actually made, but by a series of switches that operated on the genes themselves. The idea was that a particular protein began to be made when the gene specifying it was switched on and that it ceased to be made when the gene was switched off. (A gene is switched on when RNA is transcribed from it, switched off when there is no such transcription.) This model demanded that the RNA that was transcribed from the gene and transferred to the cytoplasm to serve as a template for the synthesis of a particular protein had to be short-lived. For if the template persisted in the cytoplasm as a functional structure for a long time after the gene had been switched off, then switching off the gene would not in itself stop the synthesis of the corresponding protein. If the RNA that carried the genetic instructions was stable, then a model based solely on switches that operated on the genes themselves (the genetic operator model) was untenable.

Now it was firmly established that the RNA in the ribosomes was very stable indeed, so that if the genetic operator model advanced by Jacob and Monod was correct, then the RNA that carried the genetic instructions for the synthesis of proteins could not be the ribosomal RNA. Jacob and Monod proposed that the genetic instructions were carried by a very small fraction of the cellular RNA which had quite different properties and for which they proposed the name 'messenger' RNA. The cardinal identifying property of messenger RNA was said to be that it had a short life in the cell (Fig. 1).

The experiments on which the genetic operator model was based had all been done with the colon bacillus, *Escherichia coli*, but, as might be expected from Monod, the conclusions were generalized. It was argued that the synthesis of proteins in animal cells would likewise be regulated by gene switches and that the messenger RNA in these cells would also be short-lived. These ideas were enormously influential. Indeed, even before the paper by Jacob and Monod had appeared in the *Journal of Molecular Biology*, two papers had already been published in *Nature* in which it was claimed that the short-lived messenger RNA postulated by Jacob and Monod had been identified. These two swallows did make a summer, for they were soon followed by a swarm of papers in which short-lived messenger RNA was identified wherever it was sought.

It was in this atmosphere of high excitement that we began our next series of experiments using the new method we had devised for separating the nuclear from the cytoplasmic RNA. Once again we monitored the passage of the radioactive label through the RNA of the cell and once again we reached the same conclusion: a large part of the RNA that was synthesized in the cell nucleus broke down almost as soon as it was made. But there was no RNA like that in the cell cytoplasm. I then decided to have a look at what happened to the labelled RNA when the nuclei that contained it were incubated without their cytoplasm at body temperature. I expected that the labelled RNA would break down, of course, but I wanted to examine

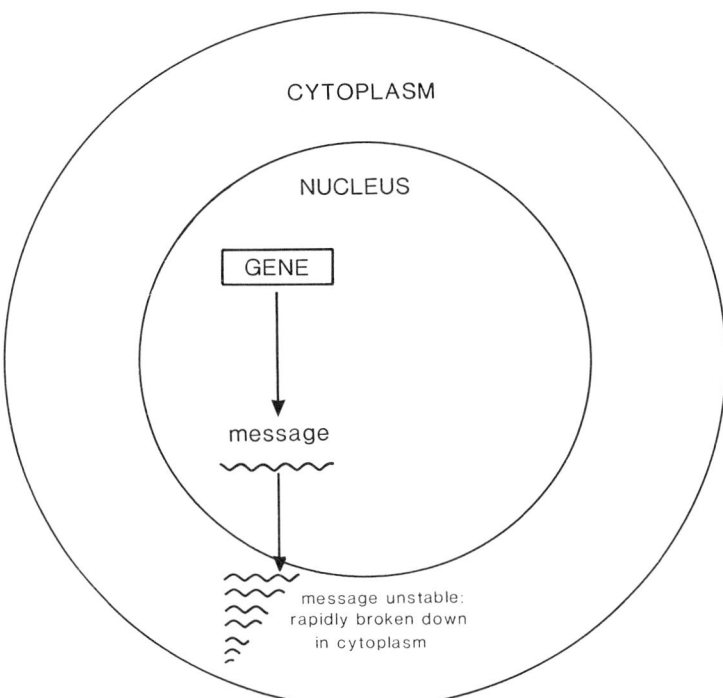

Fig. 1. Monod's view of the cell. The message that the gene transmits to the cytoplasm is highly unstable, and the messenger RNA is rapidly broken down in the cell cytoplasm.

the products of the breakdown. If, in the intact cell, the breakdown of the RNA took place within the nucleus, it would be reasonable to expect that it would yield products that could be reutilized, for it seemed very unlikely that the cell would make large quantities of RNA and then degrade it at once into fragments that could not be used again.

It turned out that when the RNA in the isolated nuclei broke down, the reaction yielded precisely those primary building blocks that were required for RNA synthesis; but when the RNA in the cytoplasmic fraction was examined in the same way, the end-products of the breakdown were heterogeneous fragments that could not be directly reutilized. The circle was thus closed. Unless there was something fundamentally wrong with our methodology, we had to conclude that the short-lived RNA that we were studying was made and broken down within the cell nucleus. It therefore couldn't be the short-lived messenger that Jacob and Monod has postulated. And though we looked very hard at our cytoplasmic fractions, we couldn't find any short-lived messenger RNA there either.

Many years later, when François Jacob was my guest in Oxford, he asked me why, when I had the best evidence in the world for an RNA that was rapidly made and rapidly broken down in the cell, I was so resistant to the idea of a short-lived messenger RNA. There were two reasons. The first was a consequence of the experiments I have just described. I couldn't find any short-lived messenger RNA in the cells I was working with, and I was convinced that the short-lived RNA that I could find was not the messenger. The second reason was more fundamental. I knew, even before I had done a single experiment on RNA, that, at least in higher cells if not in bacteria, the instructions that the genes transmitted to the cytoplasm of the cell could not be short-lived. I knew this because of Hämmerling's much underestimated, but nonetheless decisive, experiments with the giant cell *Acetabularia*.

There are several different species of *Acetabularia* that can easily be identified by differences in their shape and growth characteristics. The plant develops from a small germ-cell by producing first a stalk with rootlets and later a cap with structural features that are characteristic for each species (Plate 11). A mature plant takes many weeks to develop, and, as Hämmerling had shown, during the whole period of development it remains a single cell with a single nucleus located in one of the rootlets. By removing the nucleus, or transplanting it from one cell to another, Hämmerling had established certain fundamental facts about the relationship between the nucleus and the cytoplasm in which it resided. The most important of these were the following: (1) The characteristic shape of the cell, specific for each species, is determined by substances that pass from the nucleus to the cytoplasm. When a nucleus is transplanted from a cell of one species to an enucleated cell of a different species, the transplanted nucleus will, under appropriate conditions, impose on the cytoplasm that receives it the morphological characteristics of the species from which it was taken. (2) The growth and normal development of the cell do not require the concomitant presence of the genes that determine this development. A perfectly normal stalk

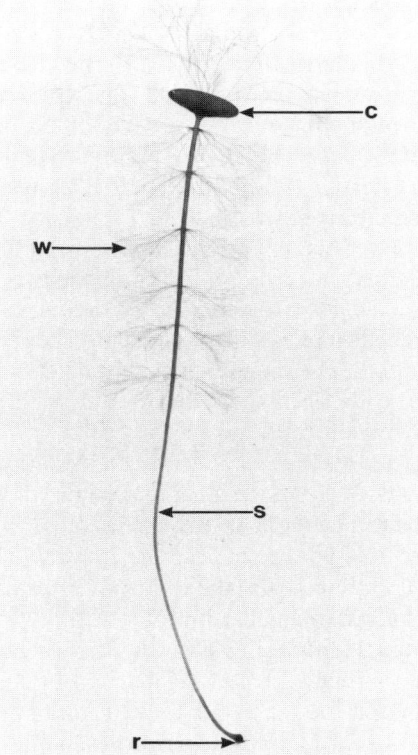

Plate 12. An *Acetabularia mediterranea* plant, showing the cap (c), stalk (s), rootlet (r), and whorls (w) which are formed before the cap develops. Although the plant is about four centimetres long, it is a single giant cell, and its nucleus resides in the rootlet. (Photograph by Professor H.-G. Schweiger.)

and cap can be generated by the *Acetabularia* cell many weeks after its nucleus has been removed. (3) The ordered sequence of changes that takes place during the development of the cell is not determined by events occurring in the cell nucleus either at the same time or immediately beforehand. All the genetic information required for the development of the plant is delivered to the cytoplasm of the cell long before it is acted upon. (4) This genetic information (the message) is very stable. Once delivered to the cytoplasm it can be maintained there in a dormant form for periods well in excess of the normal life-span of the cell and can then be brought into play by appropriate cultural conditions.

All this was diametrically opposed to the idea that the regulation of events in the cytoplasm of the cell was achieved by gene switches operating through ephemeral messengers. Because they were more direct, I found the experiments on *Acetabularia* more convincing than those on *Escherichia coli*, but even if it turned out that matters were arranged differently in the two organisms, I still thought that *Acetabularia* was a better model for the cells I worked with than the colon bacillus. *Acetabularia*, like the cells of the animal body, had a discrete nucleus, whereas bacteria did not; and in the few cases where animal cells had been examined after loss or removal of their nuclei, for example the reticulocytes that made haemoglobin, it was clear that the synthesis of specific proteins could continue in the cell long after its nucleus had gone.

Monod must have known about Hämmerling's experiments with *Acetabularia* when he made his great generalizations, but he was either unconvinced by them or chose to ignore them. I couldn't. There was, however, one feature of Hämmerling's work that could easily account for its failure to make a strong impression on molecular biologists: it lacked molecular biology. Although there was nothing equivocal about Hämmerling's findings, they had been made at a time when any attempt to reduce them to molecular terms would have been beyond the technical or conceptual resources available. I had no doubt that Hämmerling's experiments were fundamental to our understanding of how genes functioned, but I also knew that these experiments would not get the attention they deserved until they were analysed in the language of modern biochemistry. And that was what I now decided to do.

The genetic operator model had been based almost entirely on the analysis of the regulation of a single bacterial enzyme, β-galactosidase, which the colon bacillus produces when it is obliged to use lactose (sugar of milk) as its main source of energy. For *Acetabularia*, it would first be necessary to find some enzyme whose synthesis was regulated in a decisive way during the course of development; and then the key question would be whether this enzyme continued to be synthesized, and whether its synthesis continued to be regulated in the appropriate way, after the nucleus of the cell had been removed. For me, this question was actually an academic one. Since it was known that the *Acetabularia* cell could grow and produce a perfectly normal cap in the absence of the nucleus, I took it for granted that the enucleated cell was able to synthesize all the enzymes that were necessary for the production of this cap. However, it was obvious that my confidence on this point was not universally shared, and the biochemical exercise, scholastic though it might be, had to be done.

I wrote to Hämmerling who was then the Director of the Max Planck Institute for Marine Biology in Wilhelmshaven and asked him whether he would be willing to accommodate someone from my laboratory for a long enough period to acquire competence in handling *Acetabularia*, which, I understood, was a pretty tricky business. Hämmerling generously suggested a stay of about six weeks. I put it to Marianne Jahnz, who spoke German, and she agreed. A couple of months later she arrived in Wilhelmshaven and thus initiated an association between the two laboratories that continues to the present day.

My position in the world of RNA had meanwhile improved. I was no longer ignored; I was opposed. By 1961 no one was any longer surprised by the idea that the cell contained an RNA fraction that turned over rapidly. This was not because the detailed evidence that I had presented was now accepted. It was because short-lived genetic messages had become fashionable. This did not, however, help me much; for just as I had had great difficulty in convincing anyone that short-lived RNA existed, so now I had even greater difficulty in convincing anyone that the short-lived RNA wasn't the messenger, at least not in my cells. But there were signs that some biochemists were listening, even if they remained unconvinced. I received an invitation to give a talk on nuclear RNA at the Fifth International Congress of Biochemistry in Moscow where, in addition to a few days of wide-eyed tourism, I heard Marshall Nirenberg give the first instalment of his epic decipherment of the genetic code. A little later I was asked to write a review for *Progress in Nucleic Acid Research*, a series in which all the contributors were established figures in the field. This meant that in some quarters my work was being taken seriously. The case I was making would now have to be answered.

In the summer of 1962 I was invited to take part in an international colloquium organized by the Biochemical Society in Cambridge.

Francis Crick talked about the genetic code and Sydney Brenner put the case for short-lived messages. I was the last speaker. I marshalled all the evidence I had for the presence of a short-lived RNA that was made and broken down within the nucleus and for the absence of any such RNA in the cell cytoplasm; and I ended by saying that I doubted whether short-lived messengers existed in higher cells. The large auditorium was crammed with people, and when I had finished the incredulity was palpable. Francis Crick took up the cudgels on Sydney Brenner's behalf. Francis can be devastating in debate, but the back-streets of Sydney had made me pretty hard to devastate, and I think I gave as good as I got. For Francis it was all good, clean fun; but for me it was a matter of life and death.

Later in the day I had an opportunity to speak to Francis in less theatrical circumstances. I asked him how he proposed to reconcile Hämmerling's experiments on *Acetabularia* with short-lived messengers. 'What worries me about *Acetabularia*', he said, 'is all that carbohydrate'. I am not sure to the present day just what he had in mind, but his remark confirmed my view that the classical experiments on *Acetabularia* would not be taken seriously by molecular biologists until the data could be presented to them in a language with which they were familiar. It seems that Francis did not forget our gladiatorial exchange in Cambridge. In later years I twice heard him discuss intranuclear RNA turnover. By then the idea that a lot of RNA was rapidly made and rapidly broken down within the nucleus was so generally accepted that few people remembered that it had had to be fought for. However, on both occasions, Francis introduced his remarks by pointing out that it was my work that had first drawn attention to the phenomenon.

Marianne came back with the seed cultures of *Acetabularia* and the ability to grow them. This then required clean sea-water, for an adequate synthetic growth medium had not yet been devised. We arranged for large batches of deep sea-water to be collected for us from time to time by the North Sea fishing fleet and found, to our relief, that the *Acetabularia* cells grew well in it. It was not too long before we were ready to screen the cells for some enzyme that would meet our requirements. I had by this time made all the staff appointments that had been agreed: Trevor Spencer who, a few years later, emigrated to Canada; Bill Riley who ended up in Australia; Alan Rodgers who stayed on at the John Innes after I left it to return to Oxford; and Mike Bramwell who for twenty-five years has remained my close associate, mentor, and friend.

There was evidence in the *Acetabularia* literature that phosphatases might be interesting. These are enzymes that act on compounds that contain phosphate; the end result of their action is that the phosphate is split off. One such enzyme is involved in the construction of the cell wall. Trevor examined the phosphatases in *Acetabularia* in great detail and found that there was one whose synthesis increased dramatically when the cell began to form its cap. The question was: would the synthesis of this enzyme increase dramatically when the cap was formed in the absence of the nucleus?

It took us several months to settle the question, but the answer was unequivocal. When the cap began to form in the enucleated cell, synthesis of the phosphatase increased sharply just as it did when the nucleus was present. This meant that the template (the messenger) for the phosphatase was perfectly stable in the cytoplasm of the cell and that the synthesis of the enzyme on this template was regulated by mechanisms that operated in the cytoplasm and not in the nucleus. The genetic operator model, which proposed that the regulation of protein synthesis was governed solely by gene switches that worked through unstable, short-lived messengers, was clearly inapplicable. We published the work in the *Biochemical Journal* where it could hardly avoid being noticed; but it did not deflect a small army of enthusiastic young biochemists from wasting their time in a sterile search for short-lived messengers in higher cells. The consensus of opinion was that *Acetabularia* was a special case.

Acetabularia posed another important problem for us. The more closely we analysed the incorporation of radioactive precursors into RNA, the clearer it became that the relationship between nuclear and cytoplasmic RNA was very complex. While it could hardly be doubted that some RNA moved from the nucleus to the cytoplasm, none of the experiments that had so far been done excluded the possibility that some RNA might also be synthesized in the cytoplasm of the cell. This was an extreme heresy, for the 'central dogma', as it was called, of molecular biology was that DNA made RNA and RNA made protein. Since it was assumed that all the DNA in the cell was confined to the cell nucleus, the proposition that synthesis of RNA might take place in the cytoplasm was almost an affront.

None the less, Schweiger and Bremer, working in Hämmerling's laboratory, had provided decisive evidence that RNA *was* synthesized in the cytoplasm of *Acetabularia* cells from which the nucleus had been removed. Needless to say, this finding, like everything else about the behaviour of *Acetabularia*, was largely ignored or, worse still, disbelieved. Hans Schweiger spent a couple of days with us at the John Innes while we were immersed in the *Acetabularia* work. I found that his attitude to the 'central dogma', or any other kind of dogma, was much like my own, and I formed the impression that anything that this solid and meticulous scientist did had a very high chance of being right. Hans and I became life-long friends, and when Hämmerling retired I was delighted to learn that Hans was appointed as his successor. I entirely accepted his evidence for the synthesis of RNA in the cytoplasm of *Acetabularia*, and I didn't at all regard *Acetabularia* as a special case. It was important, none the less, to see whether this synthesis could be demonstrated in any other cell.

This was not so easy. No other cell could be enucleated as simply as *Acetabularia* or with so little perturbation of normal cell behaviour. In other cells, removal of the nucleus, or its natural elimination as in the case of the reticulocyte, left a residue that was severely impaired. If one found RNA synthesis in the residual cytoplasm of such cells, that was decisive, but a negative result meant very little. There were compounds that could inhibit the transcription of DNA into RNA, but they had to be used in high concentrations and were very toxic. A negative result with these was also unconvincing. What was wanted was some situation or technique that abolished the production of RNA in the cell nucleus without producing drastic side-effects in the cell cytoplasm.

It occurred to me that nature had already presented us with just such a situation. When a cell divides into two there is a stage in which its chromosomes are screwed down into portable packages for symmetrical distribution into the two daughter cells. When screwed down in this way the chromosomes are biologically inert and do not support the synthesis of any RNA. So the key question was whether any RNA was synthesized elsewhere in the cell during the stage when the chromosomes were inert. For the animal cells I worked with, this stage was too short to permit the appropriate experiments to be done. However, in some plant cells the chromosomes remained inert for several hours, which would certainly be long enough if there were no other complications. I talked the problem over with Len Lacour, who was one of the John Innes botanists who had been transferred to the Department of Cell Biology, and he suggested the growing root tips of the broad bean, *Vicia faba*, material with which he was very familiar.

Len Lacour was a remarkable man. His formal education had been truncated by the Great Depression and he had joined the John Innes as a trainee gardener. From that modest beginning he had gradually worked his way up to become an acknowledged authority on the behaviour of plant chromosomes. He was eventually elected a Fellow of the Royal Society, one of the very few in modern times to have received that distinction without the benefit of a university education. We pooled our skills and made a thorough study of RNA synthesis during that period when the chromosomes of the cells in the root tip were inert. The results were decisive. Synthesis of RNA continued in the cytoplasm of the cell during that period when synthesis of RNA in the chromosomes was suspended. So it appeared that what was true for *Acetabularia* was also true for broad beans and was hence likely to be true for plant cells generally. We published this work in *Nature* in the hope that it would attract attention, but it received no better treatment than Hans Schweiger's demonstration of cytoplasmic RNA systhesis in *Acetabularia*. Cytoplasmic RNA synthesis, like intranuclear RNA turnover, had the severe disadvantage that there was no place for it in the model of the cell that was then fashionable.

The cytoplasm of plant cells contains a set of small structures called chloroplasts which house the machinery that releases energy from the light that reaches the plant; and the cytoplasm of both animal and plant cells contains another set of small structures called mitochondria that release energy from other sources. A few years after Len Lacour and I had published our paper, it was shown that both chloroplasts and mitochondria contain DNA, that they have their own genes, and that these genes are both expressed and replicated. It was then accepted as a matter of course that these structures made RNA; and the question was no longer whether RNA synthesis took place in the cell cytoplasm, but what the function of this RNA might be. By that time, however, Hans Schweiger's pioneering work had been all but forgotten.

One bright summer morning a letter arrived from Florey. It appears that he was not entirely content with my having wandered off into the plant world and had suggested my name to the electoral board for the Chair of Pathology in Cambridge. This had at long last been vacated by H. R. Dean who, having been appointed under old statutes that set no mandatory retiring age, had remained in post until his eighty-second year. By all accounts, the department had been reduced to a pretty sorry spectacle. Florey's letter is a perfect example of his epistolary style. It is worth reproducing:

Dear Harris,
    I have been asked to obtain from you a Curriculum Vitae and a list of your publications and your present research interests. This is in connection with the Chair of Pathology at Cambridge. I wish to be quite clear that it is very unlikely that you would receive an invitation to the Chair but your name went into the hat to-day and I have been asked to provide the Board with more information.
    I may say for your guidance that the job promises to be very unpleasant in many respects. There is far too much for the Professor to do. There is great uncertainty whether the stipends that Cambridge will put up are sufficient to attract a subordinate staff and so on. There are, of course, compensating advantages in that Cambridge is a very scientific centre, etc.
    You may say that you do not wish to be considered in any case and if so will you let me know to that effect. Otherwise I await your information which will be forwarded to the proper authorities in Cambridge.
                      Kind regards,
                            Yours sincerely,

                                H. W. Florey

    The Pathology Department at Cambridge was obviously a very poor exchange for Bayfordbury, but I was flattered to be thought a plausible candidate for a Cambridge chair at so young an age. (In England views on what the appropriate age might be tend to be more conservative than they are in the United States.) I was also very curious to see how far matters would go. So I sent Florey the documents he wanted and allowed my name to go forward. A few weeks later, I received a visit from two of the electors, Wilson Smith, an eminent bacteriologist, and Frank Young, who was then the Professor of Biochemistry at Cambridge. I had a high time showing them what was going on in the Department of Cell Biology, and, like everyone who visited us, they were much taken with Bayfordbury. At the end of the day they told me that neither the facilities nor the carefree life I enjoyed at the John Innes could be matched in Cambridge and that I would be well advised to stay where I was. I heard no more about the Cambridge chair.

The battle for intranuclear RNA turnover began with a frontal assault on the whole idea. It appeared in a letter to *Nature*. The authors offered a generalized mathematical model for the transfer of RNA from nucleus to cytoplasm and contended that all our data could be accommodated by their model without invoking breakdown of RNA either within the nucleus or outside it. A glance at our data showed that they could not be accommodated by the model proposed, and I asked Neil Gilbert, the most powerful of the mathematicians at the John Innes, to help me frame an appropriately mathematical reply. A few months later, a revised version of this mathematical model appeared, again in *Nature*, but the revision didn't fit our data either. Another mathematical reply from me. Mathematical arguments rarely resolve biological controversies. The main achievement of these slightly polemical exchanges was to draw attention to the fact that an important issue was at stake and had to be resolved.

A more serious challenge to our work came from biochemical experiments in which a new antibiotic called actinomycin D was used. This compound has the ability to bind to DNA and prevent its transcription into RNA. If cells are exposed to high concentrations of it, synthesis of RNA in the cell nucleus is inhibited, and the fate of the RNA that has already been made can be examined under simplified conditions. I would never of my own accord have chosen to work with actinomycin D, for it is an extremely toxic compound which, at the concentrations required to inhibit nuclear RNA synthesis, kills the treated cells within a few hours. Results obtained under such traumatic conditions are inherently suspect.

However, one of the most unfortunate consequences of the sustained attack on my position was that I felt obliged to undertake experiments for the specific purpose of showing that my opponents were wrong. This is usually a mistake and rarely leads to interesting new observations. But what was being claimed for the experiments that had been done with actinomycin D was crucial: that the RNA made in the nucleus before the cells were treated with the antibiotic was not broken down but was simply transferred to the cytoplasm. So I was forced to make a thorough study of the effects of different concentrations of actinomycin D, and I was able to show convincingly that at the high concentrations required to inhibit nuclear RNA synthesis completely, virtually all the pre-formed nuclear RNA broke down. That didn't help me much either, for it was then argued that the breakdown was induced by the actinomycin D. In the end I won through, but it was a Pyrrhic victory. I had devoted almost three years to defensive, scholastic experiments of this kind, and I had made no real progress in understanding the biological significance of the phenomenon I had discovered.

One by-product of all this unprofitable activity was that I had become acutely conscious of the pitfalls that beset experiments in which radioactive precursors are used to study the behaviour of RNA in the intact cell. I looked at all such experiments with a very beady eye; and the more I looked at the two celebrated papers in which it was first claimed that short-lived messenger RNA had been identified, the more convinced I was that they were flawed. Sydney Brenner, who was a co-author of one of them, paid us a visit one day accompanied by Francis Crick. I spent a couple of hours closeted with the two of them in my office and spelt out the objections I had to Sydney's experiment. Sydney had no defence that satisfied me, but claimed that additional data were available that invalidated my criticisms. However, a few weeks later, in response to a prod from me, Francis wrote to say that my general point was conceded.

The messenger RNA hypothesis, as it was originally propounded by Jacob and Monod, had two essential strands. The first was that the RNA that carried the genetic specifications for the synthesis of specific proteins (the messenger) was not the ribosomal RNA, which made up the bulk of the RNA in the cell, but a qualitatively distinct component that was present only in small amounts. The second strand was that this minor RNA component had a very short life in the cell. This transience was not an adventitious property of messenger RNA, but an essential requirement of the genetic operator model for protein synthesis; for, as I have explained, the synthesis of proteins could not be regulated solely by switches that operated at the level of the genes unless the messengers emanating from these genes were constantly and rapidly replaced. Now my own work had convinced me that this part of the messenger RNA hypothesis could not be true for the higher cells that I worked with. It was perfectly clear that the instructions that the genes transmitted to the cytoplasm of such cells were not unstable, were not, on any biologically relevant time-scale, transient. And the only RNA I could find that did turn over rapidly in the cell was obviously not the messenger, because it was not transmitted to the cytoplasm at all, but was made and broken down within the nucleus (Fig. 2). I therefore felt, despite the

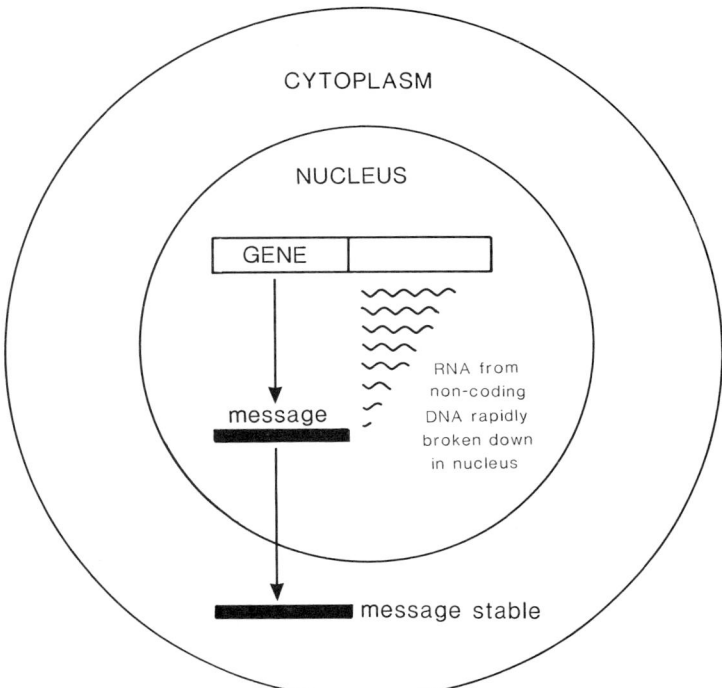

Fig. 2. My view of the cell. The message that the gene transmits to the cytoplasm is stable, but RNA made on regions of the DNA that do not code for proteins is rapidly broken down within the nucleus.

massive weight of contrary opinion, pretty secure in rejecting the idea that genetic messages had the metabolic characteristics that Jacob and Monod proposed.

However, I had no decisive evidence about the first strand of the messenger RNA hypothesis: that the genetic instructions were carried by a minor fraction of the cellular RNA that was quite distinct from the ribosomal RNA. Although circumstantial evidence in support of this idea continued to pile up, none of it was in my view conclusive. I now knew so much about the pitfalls that beset the interpretation of experiments involving the use of radioactive RNA precursors, that I had little difficulty in picking holes in each new dramatic claim as it came up. This kind of perfectionism is rarely profitable in biology. You usually have to make decisions on the balance of probabilities, and there was no doubt that the balance of probabilities favoured the view that a messenger RNA distinct from ribosomal RNA did exist, even if it did not have the properties that had been claimed for it. But science, like all other aspects of human behaviour, is only partly rational, and I remained sceptical long after the existence of messenger RNA as an entity distinct from ribosomal RNA was quite generally accepted.

Indeed, I was not fully convinced until Jean-Pierre Ebel and his colleagues in Strasbourg eventually demonstrated in the early 1970s that ribosomal RNA was chemically homogeneous. An RNA fraction that contained transcripts from a number of different genes could not be chemically homogeneous, so I finally had to accept that genetic information was transported from nucleus to cytoplasm by something other than ribosomal RNA. The moral of the story, as Thurber puts it, is that you might as well fall flat on your face as lean over too far backwards. But I did turn out to be right about the two central themes of my own research: the messages that the genes of higher cells transmit to the cytoplasm proved in the end to be perfectly stable; and the unstable RNA that I had discovered in the cell nucleus was not a genetic messenger, but had a quite different, and remarkably interesting, function that I shall discuss in a later chapter.

If my research at that time had all the aspect of static trench warfare, great changes were taking place elsewhere in the system. The most astonishing of these was that I was being rapidly transformed from a city boy into a country gentleman. Dodds was always apologetic about the word 'horticultural' in the official title of the John Innes and didn't care for the word 'institution' either. Despite the opposition of the traditionalists among the Trustees, he eventually succeeded in obtaining permission to change our name, and one fine morning we ceased to be the John Innes Horticultural Institution and became the John Innes Institute. This didn't alter the fact that a substantial part of the work remained horticultural in character, and plant breeding of various kinds was very much in evidence. There were fields planted with strawberries, others with snapdragons, and orchards that contained different varieties of apples and cherries. The glasshouses were full of experimental plants, and, for no very good reason that I could discern, a robust cow-man tended a large herd of cattle.

All this rural activity was of course quite new to me, but I found it fascinating and quickly developed an amateur's taste for it. It was, after all, so much more beautiful than my own experimental material which grew in symmetrical transparent containers stacked in closed rooms that were maintained at unpleasantly high temperatures. I also took steps to remedy my almost laughable inability to identify plants, and I was delighted to find that by the time I left the John Innes I could put a name to most of the things about me. There was nothing seriously academic in all this, but an affectionate interest in plants has remained a constant part of my life; and whenever I have a chance to spend a couple of days in a city that is new to me, I never fail to pay a visit to its botanical garden, if it has one.

The other great debt that I owe the John Innes is that my contact with the genetical work going on there was directly responsible for what was to be perhaps the most important experiment I ever did. In the summer of 1961 John Fincham, who had joined the staff of the John Innes as head of the department of genetics, returned to the Institute from the USA. He was then exploring the mechanism of what is known as complementation. It sometimes happens that when two parental stocks of a particular organism carry different genetic defects, the progeny of a cross between the two may be perfectly normal because each parent makes good, or complements, the defect in the other. Fincham was studying a special case of this phenomenon in the bread mould, *Neurospora crassa*. This mould had been introduced into genetical research by George Beadle and his colleague Ed Tatum, and it quickly became the preferred experimental material for studies on the relationship between genes and the enzymes that they specify. The problem that Fincham was exploring involved an enzyme known as glutamic dehydrogenase which acted on glutamic acid, one of the amino acids that the cell uses as building blocks for the synthesis of proteins. Fincham had strains of the mould that carried genetic defects in this enzyme. In certain cases, when the two sets of parental genes, each carrying a different defect, were brought together in a single cell, that cell behaved normally. Complementation had occurred.

In *Neurospora* and certain other fungi, the two sets of parental genes can be brought together within a single cell by a very interesting process. The branches, or hyphae, of one strain of fungus fuse with the hyphae of another, so that hybrid hyphae are formed that contain within single cells nuclei derived from both parental strains. Such hybrid cells are known as *heterokaryons*, a Greek tag meant to signify that they contain nuclei of different kinds. John Fincham's work was often brought to my attention because he needed large amounts of material for his experiments and grew batches of fungi in washing-machines which I saw from time to time being wheeled from one place to another. I became progressively more interested in the mechanism by which the fungal hyphae fused together and in the heterokaryons that this fusion generated.

Now one of the great disadvantages of somatic cells (the cells of the body apart from sperm and ovum) was that they did not lend themselves to genetic analysis. All about me people were applying genetical methods to higher plants, fungi, even bacteria, but these powerful methods were inapplicable to the cells I worked with. Genetical methods can be applied only to organisms that exchange genetic material in some way: either by sexual processes, or by cell fusion, as in the hyphae of fungi, or by direct transfer from one cell to another of the whole or a part of a chromosome, as in bacteria. But none of these processes, as far as was known, took place in somatic cells. Although, in the end, these were the cells that interested all of us most, our understanding of how they functioned lagged far behind what we knew about fungi or bacteria. It was all too obvious that anyone who could find a process in somatic cells that resulted in the exchange of genetic material between them, or who could devise a method of imposing such an exchange, would open the door on a brave new world. But how was it to be done?

This was the context in which the fusion of hyphae in fungi became increasingly interesting to me. I assumed that there couldn't be anything like that occurring naturally in somatic cells. In fungi, the exchange of genetic material achieved by the fusion of hyphae had an obviously important biological role, but it was difficult to see what purpose could be served by genetic exchanges between the somatic cells of any one individual since, at least to a first approximation, all the cells of any one individual were genetically identical. None the less, every medical student knew that there were special situations in which somatic cells did fuse together to form multinucleate cells. The classical case was in the formation of muscle during embryonic development, where single precursor cells fused together to form multinucleate tubes which were the essential mediators of muscle contraction. Moreover, pathology textbooks described a few unusual situations in which a similar mechanism was said to operate, for example in the formation of multinucleate 'giant' cells by the fusion of macrophages. But cell fusion certainly didn't seem to be a process in which the generality of somatic cells engaged, and it was very difficult to see how the few special cases that had been described could be exploited for genetic analysis. Guido Pontecorvo, who was the discoverer of the genetic exchanges mediated by the fusion of hyphae in fungi, had set himself the task of searching for similar phenomena in somatic cells, but he had not reported any success.

So it was all more or less a pipe dream until, towards the end of 1962, an issue of the *Experimental Cell Research* arrived at the John Innes. That issue contained two papers by Yoshio Okada and one by Okada and Jun Takodoro describing a detailed study they had made of the fusion induced in suspensions of cancer cells by the action of a virus that they called HVJ, an acronym for 'haemagglutinating virus of Japan'. In the pathology books there were descriptions of multinucleate cells in the lesions produced by certain viruses, and, in a vague way, I was aware of this. But I made no connection between this background knowledge and my interest in the genetical analysis of somatic cells until I read the papers by Okada and Takodoro. Then the penny dropped. I scampered back through the literature and found that there were several viruses that induced the formation of multinucleate cells, both in the infected host and, of greater importance to me, in populations of cells grown outside the body. However, in none of the earlier examples that had been described was the fusion reaction so rapid or so well controlled as that reported by Okada.

Moreover, in the paper by Okada and Takadoro, it was shown that if the virus was treated with appropriate doses of ultra-violet light, its ability to infect cells was destroyed, but its ability to fuse them together remained unimpaired. This meant that preparations of inactivated virus could be used to promote cell fusion, which would avoid the serious complications that might result from the use of infective virus. It was now crystal clear to me that the thing to do was to use inactivated HVJ virus to induce fusion between *different* kinds of somatic cells and then see whether viable heterokaryons were formed. I have the habit, and had it then, of jotting down experimental ideas that come into my head, and I wrote a note about inactivated HVJ in the notebook that I kept for this purpose. I also wrote to Okada for reprints which arrived in due course. But, astonishing as it may seem in retrospect, I did nothing more about it.

There were two reasons for this. The first was that I was not myself familiar with virological techniques, and there was no virologist on the staff of the John Innes. This meant that the routine methods for growing, collecting, and standardizing viruses were not available to me. This was not, however, a serious impediment, for the National Institute of Medical Research at Mill Hill was less than an hour's drive away, and I could certainly have obtained a supply of virus from there had I made the effort. The second reason was more fundamental. I was too deeply immersed in the study of RNA to consider embarking on a completely new, and highly speculative, experimental programme. The battle for intranuclear RNA turnover had to be won before I was ready to do anything about cell fusion. But thoughts about somatic cell heterokaryons continued to float through my mind.

There's a worm at the heart of present happiness. One morning Dodds called me over to his office and began to talk about generalities in a hesitant and slightly embarrassed way that was quite uncharacteristic of him. It finally transpired that discussions were taking place in the Agricultural Research Council about the possibility of transferring the John Innes to Norwich. This was the time when, in the wake of the Robbins Report, new universities were springing up all over the country, and one of them, the University of East Anglia, had been established at Norwich. The criteria that guided the location of these new foundations remain interesting. In almost every case, an old and picturesque city was chosen, preferably with a mediaeval centre containing a cathedral or a castle, and almost invariably at some distance from the major centres of population. The spell of Oxford and Cambridge was apparently irresistible, and it is astonishing that the planners, a group of reputedly practical men, should have thought it a sensible national policy for higher education to set up pale shadows of the ancient residential universities in out-of-the-way places.

The connection between the University of East Anglia and the Agricultural Research Council was a botanist called Bennet-Clark who was the first Dean of Biological Sciences in the new university and at the same time a member of the Agricultural Research Council. He seems to have been able to convince his colleagues on the Council that the transposition of the John Innes Institute to a site adjacent to the University of East Anglia would be of great benefit to both. The benefit to the John Innes was said to be that it would have a close association with a university instead of being isolated in the depths of the Hertfordshire countryside.

Dodds asked me what I thought of the idea. I told him that I thought it was preposterous. It was based on the same old primitive philosophy that you could establish universities and scientific institutes in the same way as you built supermarkets. In my view there was no university in Norwich worth being associated with. Even if everything went according to plan, which was unlikely, twenty or thirty years would have to go by before anything that I would recognize as a university could possibly be functioning there. Moreover, although the John Innes was located in a rural setting (immense blessing), it was midway between London and Cambridge and not more than an hour's drive from either. It was certainly much less isolated where it was than it would be in Norwich. But my most fundamental objection concerned people, a commodity that rarely seems to be uppermost in the minds of scientific planners. It was a time when skilled scientists were much in demand, and it could certainly not be assumed that the key people at the John Innes would agree to go to Norwich. Indeed, I thought that most of them wouldn't, and I was sure that if the plan went ahead against the wishes of the scientists themselves, the John Innes would simply be destroyed. I told Dodds categorically that if the institute moved to East Anglia, I would not be moving with it.

Although I was initially assured that no decision had yet been made, and that the views of the working scientists would naturally be taken into account, the administrative machinery ground on relentlessly, and it was not long before it became apparent that the Institute was going to be moved whatever the scientists thought. As I predicted, very few of them were enthusiastic about the idea, and many of them were so unenthusiastic that long before the move actually took place, they had found positions elsewhere. Not a single head of department moved with the Institute, and Dodds himself left the country altogether for a position overseas. The Bayfordbury estate was taken over as an annexe for a nearby technical college which did not seem to have the funds, or skill, or perhaps even the desire to maintain it. When, a few years later, curiosity took me back there again, I found the gardens neglected, the old mansion in disrepair, the rose collection gone, and the beautiful new Department of Cell Biology reduced to the shabbiness of an ill-used classroom. I have never gone back.

While my small world was thus crumbling about me, Florey resigned his Chair to become the Provost of the Queen's College at Oxford. Although for me that particular chair had a great sentimental attraction, I am not sure that it would have been enough to move me from Bayfordbury if the John Innes had not been doomed to destruction. But in the circumstances that then prevailed, I of course put in an application. Nothing happened for several months, and I heard all sorts of rumours, hardly worth recounting as each one was contradicted almost at once by the next. I assumed that I was not wanted. I had now reached the stage of looking regularly at the columns of academic vacancies that appeared in *The Times* and, early in the new year of 1963, noticed an advertisement for a chair of chemical pathology in the Postgraduate Medical School of London. I put in for that too and, on April Fool's day, received a letter informing me that I was not wanted there either. I was still searching the columns of academic vacancies when, on 1st June, a letter finally arrived from Oxford. It was a model of brevity from Folliott Sandford, the Registrar:

Dear Dr Harris,
   I am writing, on the instructions of the Electors to the Professorship of Pathology, to say that they have today elected you to the Professorship.
   I should be grateful if you would confirm that you are willing to accept the appointment and if so when it would be convenient for you to take it up.
                              Yours sincerely,

                                        F. H. Sandford

Some years later, J. L. Harley, the Professor of Forest Science at Oxford and a good friend, told me that when the appointment was announced, one of his colleages had asked: 'Who is this obscure botanist they have appointed to succeed Florey?'

8

RETURN TO OXFORD

A few days after the election had been reported in the *The Times* and eyebrows had returned to their normal level, I rang Peggy Turner at the Dunn School and learned that Florey had already vacated the Professorial suite. I suggested that it might be useful if I came over to Oxford pretty soon to see what needed to be done to enable me to get my experiments going. Since my requirements were very different from Florey's, I thought it likely that some modifications would have to be made over the summer if the laboratory was to be ready by the beginning of October, which I had proposed as a starting date. We settled on a day in the following week. As I drove along the familiar tortuous road from Bayfordbury to Oxford my mind was not, however, preoccupied with future plans. I kept thinking of the years gone by, of the night walks along the Sydney water-front, my Chaplinesque undergraduate experiments, the first meeting with Florey in Hugh Ward's room, the blue splash of *Chionodoxas* that greeted my arrival at the Dunn School, the explosive confrontation in the Professor's study, and Florey's subsequent generosity. And I decided, as the tower of Magdalen College came into view, that I wanted a few minutes alone in the Professor's domain before I began to take it apart. I think I wished to fix the past indelibly in my mind before it was silted over with my own activities. Whatever my motive might have been, I parked my car in an unobtrusive place and slunk into the Dunn School by a side door that led into the basement. A flight of stairs went up from there directly to the Professor's study. In the top corridor a little old man was still polishing the brass door-handles, but there was no one else, and I managed to dive into the Professor's study without being seen.

It had been stripped bare. Apart from the grey tomes containing the Collected Papers of the Department, the bookshelves were empty. An out-dated notice remained pinned to the door that led into the laboratory. The great oak desk was covered with a fine layer of dust, and the blue velvet curtains that framed the French windows were frayed at the edges. I noticed that the leather of the armchairs was crazed by a network of fine cracks. I went over to the windows and looked out at the elms which, in the summer, shut out the view of the University parks, and then came back to the desk. Two deep grooves had been gouged out of its surface and had been plugged with wax. As I examined them more closely, it struck me how dark the room was, even in June, and I at last understood why Florey always worked, not at the desk, but by the window, with his papers propped up on his knee. The drawers on one side of the desk were empty, but on the other, less used, there were a few pieces of pre-war notepaper on which orders for small amounts of laboratory chemicals had been written out in Florey's hand.

I knew that behind the panelling on one side of the room there was a bed. When he had had to vacate the house in Parks Road, Florey lived in the study for many months while he waited for his new home in Old Marston to be completed, and I had often seen the bed, partly made up, when I went into the study to speak to him. It was still there, folded up behind the panels, usable but very dusty. I tried the gas-fire. It worked, but the mantle was cracked. For use at the desk there was a slightly Dickensian, plain wooden armchair mounted on castors. I tried it. The wrong height and hopelessly uncomfortable. It was clear that although Florey had inhabited the room for twenty-seven years, he had taken no steps to make it comfortable or cheerful or even efficient. It was for him simply a *machine à habiter* and he left it as he found it. For me it was something different. Can an atheist have a shrine?

I walked through the communicating door into the laboratory. It was less gloomy than the study for it had large windows on two sides, but again there was little to indicate what had gone on there. The only piece of equipment to be seen was a huge ultra-modern photomicroscope that dwarfed a small table in the centre of the room; the surrounding benches were bare. A few bottles containing simple chemicals and stains stood on the shelves, and in the cupboards I found an assortment of glass tubes, rubber bungs, microscope slides, bric-à-brac. What a difference there already was between Florey's science and mine! I wandered around the room, looked out of the windows, pulled open some of the drawers, fingered a few bottles, and finally came to the solid oak door in the far corner. This, I knew, led to the most secret domain within the Dunn School: the Professorial bathroom suite. Only the Professor and the cleaners entered there. I opened the door and was confronted with an amazing assemblage of pre-war masterpieces: a huge bath, perhaps eight feet long with monstrous brass taps, a wash-basin that matched it in style and scale, and, through another oak door, the corresponding water-closet. Had the Oxford Preservation Trust known about it, they would surely have been tempted to seek a preservation order. I had other plans.

I knocked on Miss Poynton's door and found her in. Polite congratulations, but no chink in the formality. She must have been very close to the age of retirement, if not past it, and I asked her what her plans were. Yes, she intended to retire, but was willing to stay on, if I wished, until I had settled in and found a new secretary. I thanked her for that. Then I went down to Peggy Turner's office by way of the great oak staircase which was now dominated by a large, but ugly, portrait of Florey. Excited congratulations from Peggy and the girls in the office, followed by a large helping of the latest inside information. I asked Peggy whether she could get hold of Harry Stroud, who was now in charge of the workshop, and come up to the Professorial suite with him.

A few minutes later, I walked them both round the study and then the laboratory, and I explained what I wanted done. For the study, the main thing, it seemed to me, was to dissipate the gloom and to make the desk usable by day without artificial light. I thought this might be achieved by inserting a large skylight into the roof above the desk. Harry agreed that this could be done, but it would be no small job as the roof of the Dunn School consisted of fifteen inches of reinforced concrete. Neither of us, however, could think of any other solution. I also wanted a chair that I could use at the desk, and I proposed that, at a later stage, when the redecoration was finished, the dark, and to me unattractive, carpets and curtains could be changed for something brighter and more serene. (In the event, perhaps still under the influence of Bayfordbury, I chose pale green.)

In the laboratory I committed sacrilege. Australians dislike baths and do not use them unless there are no showers. I judged that the area occupied by this particular model would easily accommodate two service rooms and still leave enough space for a shower cubicle. I therefore suggested to Harry that the bath be removed and replaced by two small rooms, one to be maintained at body temperature ($37°$ C), which was optimal for the growth of somatic cells, and one to be maintained at $4°$ C, which I would use for storing perishable materials and for chemical manipulations that needed to be done at low temperatures. We agreed on the most suitable place for the shower, intransigent symbol of my Australian past. Alan Woodin, who was working at the Dunn School at the time, took the displaced bath lovingly into his home. There was one more thing. Florey had installed traffic lights at the door to the study in order to indicate whether he was accessible or not. They were almost always set on red. I asked Harry to remove them. I did not particularly want to see anyone else that day and left by the side door as I had come.

Congratulatory letters began to flutter into the John Innes, most of them merely formal, but a welcome few showing genuine pleasure. There was one from Florey inviting me to dinner in the Provost's Lodgings at Queen's, one from Krebs saying that he looked forward to having me as a colleague, and one, perhaps the warmest of all, from Michael Abercrombie. Michael was a man of rare charm and modesty whose stature as a scientist was not really appreciated until after his death. He had once worked in the Dunn School and was delighted that it had escaped the clutches of the clinical pathologists for another generation.

Florey's invitation to dinner was not wholly disinterested. At Oxford a retiring professor normally moves to another department if he wishes to carry on with his experiments. Florey's research was by then very much part-time and consisted of rather uninspired examination of blood vessels with the electron microscope. However, he genuinely enjoyed experimental work and didn't want to give it up altogether. He had been given house-room in the Department of Physiology, but he wanted my permission to continue to use the electron microscope in the Dunn School. It was a hilarious reversal of roles. I assured him that his presence in the Dunn School would always be welcome. He was obviously very pleased that I had been chosen to succeed him but did his best to conceal the fact. I asked him whether he had any advice to give me about the position he had held for a quarter of a century. His reply was characteristically to the point: 'The job's a cake-walk, Harris. Just don't get bogged down in administration'.

In the end the job did indeed turn out to be a cake-walk and I didn't ever get bogged down in administration; but to begin with there were serious difficulties to be overcome before I could hope to have at Oxford even a pale shadow of what I had at Bayfordbury. The reputation of Oxford as an academic institution, as opposed to its wider reputation as a picturesque back-drop to aristocratic undergraduate excess, rests on the quality of its research. However, provision for research does not loom large among the preoccupations of the central University administration or of the governing bodies of the individual colleges. The allocation of posts and the distribution of funds are determined almost entirely by the needs of undergraduate teaching. Mark Pattison's idea that the primary function of a university is to promote research has never struck roots deep enough to influence practical policy. When I asked whether the University might find it possible to provide two or three posts specifically for the support of research, I met with a very cool response. If I wanted that sort of support I would clearly have to seek it elsewhere.

My first port of call was the Medical Research Council where Harold Himsworth was then the Secretary. I made an appointment to see him without indicating that I had in mind a specific request for financial support, for I was uncertain what his reaction to my appointment might have been. I had met and admired his predecessor, Edward Mellanby, whose views on the support of medical research I instinctively shared. Mellanby's guiding principle was that he supported people, not projects. He had a profound faith in the practical importance of fundamental laboratory investigation, but clinical research he regarded as a very expensive and, on the whole, unrewarding exercise. Himsworth, on the other hand, being himself a clinician, was devoted to the support of clinical research. During his tenure of office in the Medical Research Council, the Clinical Research Centre was built in London and numerous research units were established in association with clinical chairs.

I knew that relations between Himsworth and Florey had at times been strained. The position of the Dunn School as a department of pathology was certainly anomalous. Like all the other pre-clinical departments, it had been established within the University science area long before there was any clinical school at Oxford. It had no clinical affiliation and it did not offer any pathological service for the Oxford hospitals. Neither Dreyer, the first holder of the Dunn School Chair, nor Florey, who succeeded him, had been a professionally accredited pathologist. I thought it possible that Himsworth might have had misgivings about the Chair going once again to someone without formal qualifications in pathology and even further removed from the clinic than Florey (in fact, a botanist *manqué*).

After the customary exchange of pleasantries, I asked him how he thought the work at the Dunn School might best be developed. 'The trouble with the Dunn School', he said, 'is that it speaks pathology with an accent'. By this he meant that the absence of clinical application and the paucity of accredited pathologists in the Department placed it in an eccentric and, in his view, disadvantageous position. His advice was that I should at once establish the closest possible clinical links so as to bring the Department back into the mainstream of academic pathology. My own view was that the cardinal defect of conventional departments of pathology was that they had not acquired the accent of the Dunn School. The links that I thought needed forging were not clinical affiliations but interactions with chemistry, biochemistry, genetics, microbiology, biophysics — the disciplines with which the twentieth-century revolution in biology was being made.

It was hardly a meeting of minds. What I really wanted was not advice but long-term support for a small group, so that I could bring to Oxford some of the people whom I had collected at Bayfordbury and continue the work that had been initiated there without too much disruption. However, I found myself so out of sympathy with what Himsworth was saying that I did not put the proposition to him. I saw no point in pressing my own point of view, but, as I left, I could not refrain from saying that if it was indeed true that the Dunn School spoke pathology with an accent, then I thought it very likely that by the time I had finished with it, it would be speaking an unrecognizable language.

Although my overt opposition to the transfer of the John Innes to East Anglia had placed some strain on my relations with the Agricultural Research Council, I thought none the less that the feasibility of obtaining at least some temporary support from that quarter was worth exploring. Since the plans that the Agricultural Research Council now had for the John Innes were very different from what had been presented to me when I was invited to accept the position at Bayfordbury, I thought it possible that the council might still acknowledge that it had some moral obligation to support my work, at least for a transitional period until alternative arrangements could be made. I made an appointment to see Gordon Cox, who had now succeeded William Slater as Secretary, and put my case to him. He reproached me ever so gently for my intransigence about the John Innes, but agreed none the less to provide short-term support for three scientists at Oxford. I thought this was very decent of him, for, in my experience, the agencies of government are not often oversensitive to moral obligations. When, a few years later, he asked me to serve on the Agricultural Research Council, I was glad to help.

Cox's offer, however, although very welcome, was no more than a stopgap. What I needed was a vote of confidence: stable long-term support for the person, not the project. I got it from the British Empire Cancer Campaign. That enlightened body took the view that fundamental research into the biology of the cell was, by definition, relevant to cancer and that it had to be supported on a long-term basis if it was to be any good. In response to my application, the Campaign agreed, without fuss, to establish a cell biology unit at Oxford under my honorary direction. That unit is still alive and well and has been generously supported by the Campaign for almost a quarter of a century. With the dissolution of the British Empire, the British Empire Cancer Campaign was renamed The Cancer Research Campaign, and to this splendid organization I now raise my hat.

We had a great stroke of luck when we began to look for a house in Oxford. Alexandra and I had driven over to see a house about which one of the estate agents had sent us particulars. We arrived to find that it had already been sold, but learnt on enquiry that it had been bought by an elderly couple with a well-known Oxford surname, Gee, who were vacating a larger house nearby. We called on the Gees and found that the house they were vacating was for sale, and that the plot of land on which it stood was large enough to accommodate a second home for which planning permission in principle had already been obtained. Alexandra remembered that Jim Gowans was also looking for a house. We got in touch with Jim and suggested that he might look at the Gee's property with a view to us purchasing it jointly. The idea was that when a second house had been erected, his family could live in one and ours in the other. Jim, as a good Englishman, didn't want to have anything to do with a new house, but liked the one that was already there. I, still in this respect an Australian, was attracted to the idea of building something new.

So we bought the property and divided it up between us. By a fortunate coincidence, Jim had already arranged to spend the coming academic year in New York and kindly offered us the use of the existing house while our new house was being built on the adjoining plot. What we built was something closer to a Swiss chalet than an English country house, but visiting Australians claimed (I believe in error) that there was something Australian about it. We did plant a couple of gum trees, but they didn't thrive. In this house, 73 Cumnor Hill, our children grew up, and although they have all now gone their various ways, it remains their home and will remain ours so long as we can manage it. *Invenimus portum.*

At the John Innes we began to make ready for the move. Mike Bramwell, Trevor Spencer and Bill Riley agreed to come with me to Oxford to become the founding members of the Cancer Research Campaign Cell Biology Unit, but after eight years of happy collaboration I lost Marianne Jahnz—in the nicest possible way: she went off to get married. Her first child was a girl and she called it Alexandra. Setting up shop in Oxford wasn't too difficult for Mike, Bill, and myself, but Trevor had to re-install the great glass sea-water tanks and get the *Acetabularia* cultures going again virtually from scratch. It would certainly have struck most academic pathologists as incongruous that there should be cultures of seaweed in a department of pathology. But for me their presence was a source of special pleasure. They were one of the secret weapons with which I proposed to turn pathology into an unrecognizable language. A few years later, Hans Schweiger sent me a beautiful photograph of the root system of an *Acetabularia* plant. It still adorns a wall in our common room.

There was a sentimental farewell party at the John Innes to mark our departure, but by far the most gratifying thing that happened during our last weeks there was the arrival of a letter from Howard Hiatt who was then working at the Beth Israel Hospital in Boston. He had been examining the behaviour of RNA in the nuclei of liver cells, and he wrote to tell me that his experiments also showed that the short-lived RNA was broken down within the nucleus. What was more, he had found, as we had done, that the breakdown within the nucleus yielded products that could be reutilized without modification for the further synthesis of RNA, whereas any breakdown that took place in the cell cytoplasm yielded products that could not be directly reutilized. Water in the desert.

When we moved back to Oxford we became a two-car family. To the still sleek black Morris we added a small second-hand white Renault. This was made necessary by the location of our house outside the city limits, but we could ill afford it. With the biggest mortgage that we could obtain, and all our savings committed to the new house, we were hardly in a position to embark on the high-life. But, with Alexandra's miraculous housekeeping, we somehow got over the hump, and it was not long before inflation took care of the debt. Our children also made a contribution to our standard of living. By great good fortune they all had the ability to perform creditably at examinations. In England in those days that entitled them to a good education at negligible cost.

When the work on the skylight was finished and I finally took possession of the Professor's study, I hung three photographs on the bare walls. The first had come with me from Australia. It was a shot that I had taken at Jervis Bay a few days after we were married: two skeletal gum trees silhouetted against a summer sky. The second was a romantic view of the Cell Biology building at Bayfordbury, taken from across the lake. The third was a reproduction of Dora Stock's 1789 silver point of Mozart. My reasons for having the first two are self-evident, but the third perhaps requires an explanation. It was drawn from life about two years before Mozart died and is, in my view, by far the most moving of all the likenesses we have of him. The face is just beginning to age: the eyelids are puffed, the hair a little dishevelled and touched with grey. I put it up not only because Mozart's incomparable music has been for me an unending source of delight, but also because his life is itself an inspiration, especially the last years when everything had gone — health, money, patrons, the favour of a fickle public, even the company of friends — everything, that is, except the power to create. Whenever I begin to have ideas above my station, a glance at that portrait of Mozart brings me down to earth.

But Mozart wasn't the only thing to bring me down to earth just then. There were the undergraduate lectures to be prepared. My view of undergraduate lectures was pretty radical and has remained so. As a means of imparting information — the conventional view of how things stand in a particular subject — I regarded lectures as having been rendered obsolete by the invention of the printing press. Unless a lecturer offers something more than can be found in textbooks or current reviews, I cannot see that he serves any useful function. Students can read, after all. At Oxford, no lectures are compulsory and students vote with their feet. That is as it should be. However, in order to be able to do more than rehash the potted information in textbooks, a lecturer must have given long thought to the problems he proposes to discuss. In the experimental sciences he must actually have worked on these problems. This was a principle I adopted from the very beginning, and it continues to govern all the lectures given at the Dunn School. We do not attempt to 'cover the course', and no one lectures on a subject to which he has not himself contributed.

There is one exception to this rule. The first lecture of the year is traditionally given by the Professor himself, and Florey had always chosen to give a historical introduction to the study of pathology. I never liked it. What he said was conventional, second-hand, and, as might be expected of any attempt to cover the history of pathology in one hour, very superficial. That opening lecture gave me a great deal of trouble. At first I thought we might forgo the history and substitute something else. Scientists have a very ambivalent attitude to the history of science. In an experimentalist, a late flowering of interest in the subject is usually taken as a sign of failing powers. One can certainly be a competent scientist without having any knowledge or, indeed, any interest in the history of one's subject; and in the ruthless competition of contemporary biological research, the young practitioners want to know what is happening now, not what happened yesterday. But this is simply to say that one can be a competent scientist (or for that matter a competent classicist) and still be a philistine. Yes, it is possible, but that does not mean it is desirable.

So I decided in the end, despite the absence of specialist knowledge, that I would persevere with the history, and I did something that any good historian would regard as disreputable. I turned the lecture into a piece of propaganda, using the history to illustrate the nobility of scientific enquiry and the indispensable part it has played in constructing a rational and humane view of the natural world. Oxford students are a pretty sceptical lot, but I did not notice a precipitous decline in numbers at my second lecture, which was about inflammation, or at any of the others I gave that year. I've been lecturing in that theatre now for almost a quarter of a century, and although I cannot pretend that my initial enthusiasm has remained undimmed, it is no small reward that the students continue to come.

My first piece of professorial research at Oxford was not a further exploration of nuclear RNA but an excursion into English land law. Behind the Dunn School, on either side of the causeway that joins the main building with the animal house, there were two plots of land on which, before the War, food was grown for the experimental animals. Since this had now become available from commercial sources, the land lay fallow, and several enterprising people in South Parks Road began to get ideas about how best it might be developed. When I arrived, I learned that two schemes were actually being discussed at some level within the University administration. The first was a proposal put forward by Cyril Darlington, the Professor of Botany, to convert the vacant land into a genetic garden. The second came from Ewart Jones, the Professor of Organic Chemistry, who thought the land could best be used to provide an additional car-park for the science area. I was not at all sure how territories were determined in South Parks Road, but I remembered having heard that there was something special about the Dunn School site because of undertakings that had been given in connection with Sir William Dunn's will. Peggy Turner dug up the relevant documents.

It transpired that the will had been the subject of a High Court Order. In its bid for the legacy, the University of Oxford had offered to use it to construct a laboratory, to be known as the Sir William Dunn School of Pathology, on a site in South Parks Road. The precise location and dimensions of the site were shown in an annexe to the High Court Order, and the University had agreed to allocate this site to the Sir William Dunn School of Pathology in perpetuity. Now perpetuity is a very long time, and when I asked Brian Simpson, who was then the Law Tutor at Lincoln College and an authority on English land law, what all this meant, he gave it as his opinion that the University was not free to do anything with the site that had been allocated except to use it for the further development of the Sir William Dunn School of Pathology. I sent a copy of the High Court Order and a précis of Brian Simpson's opinion to the University Registrar and heard no more about Cyril Darlington's genetic garden or Ewart Jones's car-park. Over the years, that High Court Order has been of inestimable value to me, especially in defending the Dunn School gardens against the relentless pressure of the motor car. Some people say that the resurrection of that document was the best piece of research I ever did.

The vacant land did not remain vacant for long. It was used, in accordance with the High Court Order, for the further development of the Dunn School. The idea came from Jim Gowans. The Dunn School was very crowded. Neither Jim nor Ted Abraham had anything like enough room to develop their work on the scale that it merited. The arrival of the Cancer Research Campaign Cell Biology

Unit made matters worse. A few months before my return to Oxford, the Medical Research Council had agreed to establish a cellular immunology research unit under Jim's honorary direction. While I was still at Bayfordbury, he had already mentioned to me that the Council would be willing to provide funds to build an extension to the Dunn School to house his new unit. I thought it preferable to make more intensive use of the small amount of land available and conceived the idea of erecting a larger building, with one floor for Jim, one for Ted, and one to provide more space for the Department as a whole. I managed to obtain funds from the Wellcome Trust for Ted's laboratories and from the University Grants Committee for the Departmental extension.

In due course a plain three-storeyed structure made of reinforced concrete slabs, a mode of construction then fashionable, made its appearance behind the elegant building in Queen Anne style. The extension did indeed solve our space problems, but it is hardly a joy to behold, and it made the Department a good deal bigger than I would have wished. A department becomes too big when special efforts have to be made to ensure that the people in it get to know each other. I'm afraid that is what has happened to the Dunn School, and I cannot see any acceptable way of reversing the process.

The public view of Oxford, that heady mixture distilled from *Zuleika Dobson* and *Brideshead Revisited*, and powerfully reinforced by the unflagging interest of the media in undergraduate affectation and donnish eccentricity, is one of the great myths of our time. The first thing I learnt about Oxford when, as a professor, I was able to view it from the inside, was that it was a passionately democratic institution. Some say democracy run riot. The University has no stable centre of executive power. There is no governing body, no senate or board of regents with political or mercantile representation. The Vice-Chancellorship, the chief executive office, is rotated at frequent and regular intervals. There are no deans (and never will be). The principal organs of University government, the Hebdomadal Council and the General Board of the Faculties, are elected bodies, and the electorate, known collectively as Congregation, includes all established members of the teaching staff. Every piece of proposed legislation is first published in the *University Gazette* and will only pass without debate if it meets no opposition. A handful of signatures is enough to signal opposition and to ensure that the proposal is publicly debated at a meeting of Congregation where its fate is decided by vote.

The central University authorities are very conscious of the power of Congregation and do not attempt to introduce legislation unless they believe that it will have the support of the majority. Even so, hardly a year goes by without some proposal being thrown out by a vote of Congregation. This democracy percolates down through the Faculty Boards to every aspect of academic life. I learnt very quickly that I could do anything I wanted at Oxford if I had the support of my colleagues, and nothing if I did not. Initially, I found the measured pace of democratic processes frustrating, but having over the years become familiar with other forms of university government, I would not now wish to see Oxford governed in any other way. Delay is a small price to pay for political independence and academic self-respect.

The Chair of Pathology at Oxford is attached to Lincoln College. At its first meeting of the academic year, the Governing Body elected me a Fellow, a process that required me to swear an oath, in Latin, that I would abide by the laudable customs of the College. Whether this bound me to do no more than abide by those customs that were laudable or whether all the customs of the College were by definition laudable, the language of the oath did not make quite clear. None the less I managed to rub along, and I began, very slowly, to acquire some understanding of what an Oxford college was about. Its primary function is not, as the media insist, to move port majestically around the high table, but to teach undergraduates. To this end the great majority of college dons devote the main effort of their working lives.

At the centre of college teaching is the tutorial at which undergraduate and tutor meet once a week for what is essentially an hour of private tuition. The undergraduate prepares an essay on the basis of guided reading, and the tutor in discussing it seeks to enlarge his pupil's understanding of the subject. At its best, this form of teaching is incomparably superior to any system of mass instruction, but it is, of course, expensive, and it is very sensitive to the quality of the tutor. College teaching can be an indelible experience, but it can also easily be a total waste of time — all the more so because college tutorials, unlike lectures, are compulsory. Professors do not take part in college teaching and often have an ambivalent attitude to it, especially the science professors. Progress in some branches of science is so rapid that any one college tutor, no matter how conscientious, can only hope to have a deep understanding of a very limited part of his subject, so that much of the tutorial instruction, although intensive, is none the less secondhand. The principle that I adopted for lectures in the Dunn School, that no one lectured on any topic to which he had not himself contributed, could never be applied to college tutorials for there are simply not enough tutors to go round.

As I had not been an Oxford undergraduate and had given only an occasional tutorial (in the days when I needed the money), I did not have any special attachment to the tutorial system and was perhaps more sceptical than most about its suitability for the teaching of modern science. There was, indeed, one respect in which I regarded the tutorial system as positively dangerous, and still do. College teaching makes great demands on the time and energy of tutorial fellows, and for an experimental scientist these didactic duties pose a real threat. It takes an iron discipline to combine a full programme of college teaching with creative science, and it appears that rather few people possess it. Serious experimental research, even when one has the necessary gifts, is an all-absorbing activity and incomparably more taxing than conscientious teaching. It is therefore not surprising that many of the promising young scientists who are appointed to college tutorial fellowships at an early age soon make teaching their primary concern and hence fail to attain the level of scientific distinction that might be expected of them. My early years in science had been protected by Florey; and I had not been in the post for long before it became clear to me that I too would have to ensure that the young experimentalists who began their careers with me did not succumb too soon to the demands, and the blandishments, of college life.

Access to college life did, however, bring me one immeasurable reward. During the dreary years that I spent grinding my way through the medical curriculum in Sydney, I did not lose contact with the friends I had made earlier, and the circle in which I moved was never limited to medical students. However, after leaving Australia, and especially at Bayfordbury, I had had few opportunities to meet people outside the scientific environment in which I worked. That is the lot of most scientists working in research institutes and universities around the world. But the fellowship of a traditional Oxford college is assembled from as wide a range of academic interests as the college can afford; and when I went into Lincoln to dine, I met not only other scientists, but classicists, historians, lawyers, linguists, musicians—gifted people from a host of different disciplines. Over the years my life has been constantly enriched by the company and, in some cases, the friendship of scholars whom I met in the common rooms of Oxford colleges, and that, it seems to me, is the greatest gift that Oxford has to bestow.

In the Dunn School I got down to work with as little fuss as possible. I did not at once attempt to initiate major reforms, although there were some that were surely necessary. However, I was not only the youngest professor in South Parks Road, I was also the youngest member of the Dunn School teaching staff, and I had the wit to see that in that position it was inadvisable to attempt to move mountains on the first day. In any case, as at Bayfordbury, for me the most important thing was to get my experiments going, and I still firmly believed that, if they went well, everything else would fall into place. In fact, I was busy in the laboratory long before the University term began. My first experiments were of a rather formal kind, done to dismiss some further ephemeral data that had appeared in *Nature*. Then I got stuck into something more serious. I now knew that the short-lived RNA was broken down within the nucleus and couldn't therefore be the template for protein synthesis in the cytoplasm. It was, however, possible that it was implicated in the synthesis of protein in some subtle way that had escaped me. I have never had any difficulty in accepting that there were more things in heaven and earth than were dreamt of in my philosophy.

I therefore decided to examine, as carefully as methods then permitted, the precise relationship between the breakdown of the short-lived RNA and the rate of protein synthesis in the cell. To do this I again used the antibiotic actinomycin D which at high concentrations blocked the synthesis of RNA on the genes. If the short-lived RNA played some important but cryptic role in the synthesis of proteins, then one would expect that its breakdown, under conditions in which it could not be replenished, would be reflected in the rate at which proteins were synthesized. Of course, in the presence of actinomycin D, the rate of synthesis was bound to fall in any case because of the toxicity of the antibiotic, but a careful analysis of the two processes of decay—the breakdown of the RNA and the fall in the rate of protein synthesis—might be able to show whether they were linked or not. I therefore blocked the synthesis of RNA with an appropriate concentration of actinomycin D, making sure that the synthesis was indeed blocked, and then simultaneously measured the rate of RNA breakdown and the decline in the rate of protein synthesis.

The results were unequivocal. The two decay processes were clearly unrelated and were described by quite different mathematical functions. In fact, almost half of the short-lived RNA had broken down before any change in the rate of protein synthesis could be observed. The function of intranuclear RNA turnover thus remained a mystery. The very scale of the process made it certain that it served some important function in the organization of the cell, but it began to look increasingly probable that the elucidation of this function would require some enlargement of the conceptual framework that was then current.

The other set of experiments I did that year were something in the way of an aside. Lee Sabath, then at the Harvard Medical School, was spending a sabbatical leave with Ted Abraham. He was studying the synthesis of the enzyme penicillinase in a bacterium known as *Bacillus cereus* which is the causative agent of one form of infectious diarrhoea. Penicillinase is a very important enzyme from the clinical point of view because its presence is responsible for the resistance of certain strains of bacteria to penicillin. The enzyme is secreted by the bacterial cell and destroys the antibiotic. In *Bacillus cereus*, the synthesis and secretion of the penicillinase is induced when the bacterium is exposed to penicillin or a related compound. However, Lee Sabath had found that the time-course and general character of the process of induction was very different from what Monod and Jacob had described for the enzyme β-galactosidase in *Escherichia coli*, the experimental material on which they based their theory of a short-lived messenger RNA.

It seemed to me that what applied to the β-galactosidase in *Escherichia coli* might not apply to the penicillinase in *Bacillus cereus*. I suggested to Lee Sabath that we might do a collaborative experiment to see whether induction of the penicillinase could take place even after all synthesis of RNA in the cell had been shut off. We again did this with actinomycin D, using a concentration that completely inhibited RNA synthesis, and were able to show that once the process of induction had been initiated, synthesis of the enzyme could begin and could continue long after all RNA synthesis had stopped. The messenger RNA for the penicillinase was clearly not short-lived. It thus appeared that a genetic operator model based on short-lived messengers was not generally applicable even in bacteria.

Slowly, the news that I had succeeded Florey to the Oxford Chair filtered back to Australia. The Australian version of the Florey legend had not yet reached the dimensions it was ultimately to attain, but, for the tiny segment of the population that took any interest in such matters, the fact that he had been succeeded by another Australian was an occasion for a little patriotic self-congratulation. A number of letters in that vein came in from Australia. There was, however, one that was altogether different. It came from Bruce Leckie, who had been my contemporary at medical school in Sydney and one of the instigators of the unforgettable farewell party that marked my departure from the Royal Prince Alfred Hospital. He was now a surgeon there, and he wrote to ask whether I could find the time to fly out to Australia to have dinner with him and some other members of our cohort. I was astonished that the memory I had left in the minds of my undergraduate contemporaries was warm enough to prompt so sentimental a gesture, and they cannot have had any idea of how deeply I was moved. Fourteen years had elapsed since I left Sydney, and in that time I had never had enough spare cash to permit the luxury of going back. If, of the cities around the world that I had come to know well, Oxford had become my second love, Sydney remained my first.

A couple of weeks later I received an open air ticket from Qantas, and, one morning early in April, I climbed into a Boeing 707 whose interior walls were decorated with bright sketches of Australian wild flowers. A monotonous ditty that had been a great hit while I was at school came floating up out of the past. It had a catchy refrain carried by the words 'I'm gonna make a sentimental journey', and throughout the whole of that interminable flight to Sydney I could not get it out of my head. I was met at the airport by a small contingent of ex-resident medical officers of my vintage and some members of my family. My father had sold No. 15 Poate Road and had retired to a small apartment beside the beach at Bronte, and it was there that I was taken. I slept for most of the first day and naturally woke up very early on the morning of the second. The sun was already up, but my parents were still asleep, and I decided to slip down to the beach without waking them. It was still too early for the pre-breakfast surfers, and there was no one else around. I lay down on the white sand, not yet warmed by the sun, watched the breakers roll in, sniffed the salt air, and regretted the world that I had lost.

The dinner was a much bigger affair than I had expected. It had been arranged in the Hotel Australia, which was then still Sydney's top hotel, and about thirty people turned up, some of them coming from distant country towns. We took up exactly where we had left off, as if nothing had happened since last we saw each other. Only our physical appearances betrayed that time had not been annihilated. Indeed, for most of them, nothing much had happened. They had had the good fortune to be able to live their lives in a more or less predictable fashion within the environment into which they had been born. They had achieved varying degrees of professional success; none of them had known hardship. Their domestic circumstances were the usual mixture of happiness and unhappiness, and when I had finished asking them about their careers and their families, there wasn't all that much to talk about. With one or two exceptions, their view of the world was still largely bounded by the city of Sydney, and the problems that dominated their conversation, although still of interest to me, were no longer my problems. It transpired that only one of the cohort was no longer alive—a shy man who had turned out to be homosexual and had committed suicide.

After the dinner there were a number of short speeches, mainly recollections of escapades in which I had been involved. They were all inaccurate, some completely apocryphal. It was brought home to me that memory is intrinsically myth-making. My appointment to Florey's Chair was regarded as a triumph for the group as a whole, and it seemed to me that they took genuine pleasure in it. They wanted to hear about life in Oxford, and I told them a little about it. We went on till the early hours of the morning drinking beer and becoming more and more sentimental. It was a great Australian occasion.

I had only a few days in Sydney and spent most of them visiting the haunts of my childhood. The city was changing very quickly. Tall buildings were springing up in the centre, and all around the marvellously romantic harbour the water's edge was being crowded out by high-rise apartments. Between Bondi and Tamarama the scruffy bush had given way to intensive urban development, No. 15 Poate Road had been replaced by a block of flats, Paddington had become fashionable, and the clattering trams had gone forever. The inward-looking, half-colonial backwater in which I had spent my youth was now hard to find; and when, ten years later, I had an opportunity to look for it again, it had disappeared altogether.

As the year wore on, I began to receive further indications that my experiments on intranuclear RNA turnover were gradually being accepted. A letter came from Jan-Erik Edström, then Professor of Histology at the Karolinska Institute in Stockholm, saying that he was able to confirm my observations. He had devised some remarkable micromethods that permitted the analysis of samples of RNA synthesized at different sites within the chromosome; and by the use of these methods, which were quite different from mine, he had also found that there was substantial breakdown of RNA within the nucleus. I heard from Guido Pontecorvo, who was then in New York, that Jim Darnell, who had initially been opposed to the idea of intranuclear turnover, had now come round. Invitations to lecture, on both sides of the Atlantic, began to arrive at frequent intervals, and I soon became an habitué of the interior of Boeing 707s. I paid my first visit to Montreal where I met Barbara McClintock who was then having a very hard time getting people to listen to the idea that genes could move around within the nucleus. I gave a lecture at the Rockefeller Institute, and everyone there but the notoriously critical Alfred Mirsky seemed to accept my data. The question had now become not whether intranuclear RNA turnover existed, but what its biological significance was.

Needless to say this question had not escaped my attention. If it was accepted that the short-lived nuclear RNA was not a template for the synthesis of protein and was not implicated in protein synthesis in some more subtle way, then there appeared to be only two other possibilities. The first was that this RNA was somehow involved in the regulation of gene activity within the nucleus — switching genes on and off or controlling the rate at which they were transcribed. The second was that it served some novel function that we hadn't thought of.

Wherever I went to lecture on short-lived RNA, I met the suggestion that it must be involved in gene regulation, but I never could muster any enthusiasm for this idea. To begin with, I couldn't see why a family of RNA molecules that regulated the activity of genes should have such a short life, or, conversely, how a short life could help in the regulation. All the evidence indicated that in animal and plant cells genes were not switched on and off at high speed. It appeared, on the contrary, that once they were switched on or off, they remained so for long periods of time. And then I couldn't see how regulation of genes by means of RNA would work in chemical terms. A molecule of RNA is simply a linear transcript of a linear gene, and, as far as was known at the time, two nucleic acid molecules (either RNA or DNA) could recognize each other only if the subunits that made up the two linear chains were in register (were homologous). It was easy to see how an RNA molecule could recognize the gene on which it was made, but how could it recognize, without error, a different gene with which it was not in register? So I turned away from gene regulation and cast about for some function that could provide a plausible explanation for the short life of the nuclear RNA. I was sure it would turn out to be a function that was not then the centre of attention.

Charles Darwin gave my subconscious a prod. I began to play with the idea that cells might constantly be trying out new evolutionary experiments. If the chromosomes contained regions of DNA that did not code for proteins or serve any other immediately essential cellular function, then those regions of the DNA could undergo and accumulate stable changes without in any way impairing cellular functions. Most stable changes (mutations) that occurred in the regions of the DNA that did code for proteins were deleterious, but on regions of DNA that did not code for proteins (or serve any other immediately essential function) natural selection could operate with impunity. Such regions would be, *par excellence*, the raw material for evolutionary change; and it was reasonable to suppose that they would be transcribed into RNA so that any changes they had undergone could be tried out in the cell. I began to entertain the notion that intranuclear RNA turnover was the biochemical manifestation of an evolutionary game of trial and error. More specifically, I became attracted to the idea that the RNA that broke down within the cell nucleus was synthesized on regions of the DNA that did not code for proteins, and that the biological role of this RNA was to permit new genetic combinations to be tried out.

I was given an unexpected opportunity to present these new ideas to a large and, in principle, interested audience and to publish them. Vernon Bryson and Henry Vogel, of the Institute of Microbiology at Rutgers University, were organizing an international symposium there on the subject 'Evolving genes and proteins'. It was to take place in the summer of 1964, and some months beforehand Henry Vogel rang me to ask whether I would be willing to contribute. I agreed. What I said at that symposium was subsequently published in book form, so that I have a documentary check on the myth-making propensity of my own memory. I rehearsed all the evidence I had in support of the conclusion that the short-lived RNA was broken down within the nucleus and could not therefore be a messenger RNA for the synthesis of protein in the cytoplasm. Then I took the plunge. There were about three hundred people in the audience, among them some of the best-known names in the molecular biology of that time, but in the desultory discussion that followed my talk no one thought my evolutionary explanation for intranuclear RNA turnover worth serious consideration. In the published version of my lecture, the final paragraph contains the definitive statement of the views I held at that time:

Only a small proportion of the RNA made in the nucleus of animal and higher plant cells serves as a template for the synthesis of protein. This RNA is characterized by its ability to assume a form which protects it from intranuclear degradation. Most of the nuclear RNA, however, is made on parts of the DNA which do not contain information for the synthesis of specific proteins. This RNA does not assume the configuration necessary for protection from degradation and is eliminated within the cell nucleus. It plays no role in the synthesis of cell protein, but serves as a background on which mutation and selection may operate to produce new templates for protein synthesis (Fig. 2, p. 155).

Thirteen years later, when the genetic code had been deciphered and new chemical methods had been devised that permitted the precise molecular structure of genes to be determined, it was discovered that genes in animal cells were much more complex than anyone had imagined. The elements that encoded specifications for the structure of specific proteins were not, as everyone had supposed, simple continuous linear sequences. It appeared that the coding elements, even for different parts of the one protein, were distributed in the chromosome as separate pieces, interspersed with regions of DNA that did not code for anything. For many genes, the amount of non-coding DNA greatly exceeded the amount of coding DNA. It soon transpired that in order to make a functional template for the synthesis of a particular protein (a messenger RNA), the RNA transcripts from the separate elements coding for different parts of the protein had to be brought together into an uninterrupted sequence; and this was achieved by highly specific reactions that spliced out and degraded the RNA transcripts from the intervening non-coding regions. Intranuclear breakdown of RNA transcribed from non-coding DNA was thus an intrinsic and essential part of the mechanism by which a functional template for the synthesis of a protein was produced.

In a piece entitled 'Why genes in pieces?', which appeared in *Nature*, Walter Gilbert argued that the biological function of the non-coding DNA regions was to be explained in evolutionary terms. They were, he suggested, 'both frozen remnants of history and the sites of future evolution'. I didn't find much to disagree with in that. Another editorial in *Nature* at about that time mentioned the fact that intranuclear RNA turnover had actually been discovered by me some fifteen years previously; but only old friends remembered that I had also come pretty close to the mark in elucidating its biological significance.

The academic year ended with what was to become almost an annual event—the family seaside holiday in France. This practice began when we discovered that we could have a much cheaper (and more pleasant) holiday on an unfashionable beach in Brittany than in any of the seaside resorts in the south of England; but as time went by we went to France because we loved it. Our first such holiday was in Lancieux on the rocky northern coast of Brittany, our next at Beg-Meil, which has two beaches. Each time we went a little further south, to the Vendée, Oléron, the Landes, and finally to the Mediterranean. We would take the car over via the Channel ferry and spend a lot of our time touring the countryside in the region where we were staying. We all spoke French, which made the holidays even more enjoyable; and if Alexandra did the talking we were not at once taken for foreigners. The Mediterranean brought sharp reminders of Australia: a dry landscape dotted with eucalypts, familiar semi-tropical flowers, warm night air, and the smell of orange blossoms. It used to be said that all good Americans go to Paris when they die. Perhaps, somewhere in Provence, there's a place like that for expatriate Australians.

9

CELL FUSION

When I got back to work after the summer holidays I decided that the time had come for me to turn my attention to cell fusion. During my time at Bayfordbury there had been only one new appointment to the staff of the Dunn School, a charming Welshman by the name of John Watkins. He was a virologist and was at the time actually studying the behaviour of cells infected with herpes virus, one of the many that could under certain circumstances induce cells to fuse together. He had at his command all the virological techniques that had not been available to me at Bayfordbury. One morning (I think it was towards the end of September), I had a long talk with John and put to him the ideas I had about using cell fusion to induce the formation of heterokaryons from different kinds of animal cells.

I gave him the three reprints I had received from Okada and asked him to see whether he could obtain the HVJ virus that Okada had used, if that seemed to him to be the best virus for the purpose. He thought it probably was and wrote to Dr H. G. Pereira, who was in charge of the virus collection at the National Institute for Medical Research, to enquire whether the HVJ virus was available. I could have done that myself two years previously but, for reasons that I have already explained, I was too preoccupied with other matters. Pereira wrote back to say that he didn't exactly have the HVJ, but offered a strain that he thought was the same thing. This was called the 'Sendai' virus after the Japanese city in which it was thought to have originated. By the middle of October John had grown up stocks of the Sendai virus, standardized the preparations, and inactivated some batches with ultra-violet light as prescribed by Okada and Tadokoro, who, as I have mentioned, had shown that appropriate doses of ultra-violet light could destroy the ability of the virus to infect cells without destroying its ability to fuse them together. We were ready to try the experiment.

The question was, what cells should we fuse? Here I made a decision that was to have profound consequences: I decided to fuse together cells from two different animal species. Why? There was one perfectly sound reason that would have been immediately appreciated by any geneticist. In order to analyse the genetic consequences of crossing one individual with another, one must have what geneticists call 'markers'. Markers are stable, genetically determined characteristics that permit the two parental individuals to be distinguished; for example, flower colour as used by Gregor Mendel in his classical experiments on peas, or eye colour in man. The trouble with somatic cells from any one animal species, and especially when they were grown outside the body, was that they showed very few characteristics that could be used as genetic markers. It was, however, obvious to anyone interested in such problems that if a successful cross could be made between cells from two different animal species then a virtually inexhaustible supply of genetic markers would be available.

Very few biologists at that time would have thought that such a cross had any chance of success. It was an era when biology was dominated by increasing knowledge of the extreme specificity of cellular interactions; by the findings of transplantation immunology which showed that grafts exchanged between two individuals, even of the same species, were rejected unless these individuals had closely similar genetic constitutions; by the set of ideas encapsulated in Peter Medawar's phrase: 'the uniqueness of the individual'. Even in the case of heterokaryons formed by fusion of hyphae in fungi, it was known that such heterokaryons would survive only if they were formed between closely related strains. If fusion by chance occurred between the hyphae of unrelated fungi, incompatibility systems came into play within the cell and these brought about the destruction of the rogue heterokaryon. No sane geneticist would then have thought it plausible to suppose that a viable entity could be produced by amalgamating two cells from different animal species. Even those who were seeking to develop genetical methods for the study of somatic cells did not apparently think the experiment worth a try.

It seems to have been my fate in science to see things differently. I regarded the exquisite sensitivity of the immunological reactions that determined individuality as a late evolutionary development, mediated by molecules that were confined entirely to the surfaces of a few specialized cells. On the inside, all animal cells looked much the same to me. I was much more impressed by the similarities than the differences. I thought the intracellular incompatibility systems that existed in fungi could only have developed because the fusion of hyphae was for fungi a normal system of genetic exchange on which natural selection could operate. It was remotely unlikely that at any point in evolutionary time the somatic cells of one species of animal would have engaged in systematic genetic exchanges with the somatic cells of another, so it seemed to me very unlikely that the somatic cells of animals would have developed intracellular incompatibility systems comparable to those that existed in fungi. I believed that if we could get around the obstacles posed by the cell surface, somatic cells from different animal species would, when fused together, accept each other.

But there was also an irrational element in my choice of cells, connected in an obscure way with my childhood admiration for the experiments of Wöhler. It had greatly pleased me that Wöhler had demolished the conceptual boundary between the animate and the inanimate world. Well, there was another conceptual boundary that appeared to me to be in need of demolition: the ancient doctrine, adopted in one form or another by many philosophers and most priests, that man was in some way categorically different from other animals. From the Middle Ages to the twentieth century, it had consistently been argued that man possessed some fundamental attributes that animals did not share: an immortal soul, reason, free will, language, what have you. To me all this was factitious. I could not see, indeed could not conceive, any attribute of man that did not have some representation elsewhere in the animal kingdom. I could not detect any natural boundary, only the conceptual boundary imposed by dogma. And I believed that any attempt to elevate man by separating him off from the rest of the living world merely served to diminish him. In the long view of evolutionary history, the differences between mice and men were trivial. In making my choice of cells I was also cocking a subconscious snook at these anthropocentric doctrines that I disliked. That was why the very first cells that we attempted to fuse together were taken from two different animal species much beloved by poets and story-tellers as symbols of contrast—mouse and man.

The two cell-types were HeLa, the line of human uterine cancer cells that I have already written about, and the Ehrlich ascites tumour, a long-established line of mouse breast cancer cells that grew well as a suspension in the mouse peritoneal cavity. Apart from being easy to handle, these cells had a special advantage for the experiment I had in mind: their nuclei looked very different under the microscope. It was, therefore, reasonable to expect that if we succeeded in fusing them into a heterokaryon, the two types of nuclei would be readily distinguishable within the one cell. I grew up the cells and, on 21st October according to my diary, gave them to John Watkins who had the inactivated virus ready. John mixed the two cell suspensions together, treated the mixture with the inactivated virus, and then distributed it into some flasks containing small flat glass discs (cover-slips) to which the cells could adhere. The cultures were maintained overnight at body temperature to permit the cells to grow, if they would, and the following morning the cover-slips with the adherent cells were taken out and stained for microscopic examination. John brought them up for me to look at.

It was obvious at a glance that the cultures contained numerous large cells with more than one nucleus, and in some of these it certainly looked as if the one cell contained nuclei derived from HeLa cells together with nuclei from Ehrlich cells. But, of course, with a dramatic experiment of this kind, appearances would convince nobody. So I incubated the HeLa cells with a specific radioactive precursor of DNA so that their nuclei became labelled, and we then fused the labelled HeLa cells with unlabelled Ehrlich cells. The idea was that if we made autoradiographs of the multinucleate cells on the cover-slips, the labelled nuclei would blacken the overlying film while the unlabelled nuclei would not. The presence of labelled and unlabelled nuclei within the one cell would be decisive evidence that we had indeed produced heterokaryons containing both human and mouse nuclei. When the autoradiographs were developed, that is exactly what we found (Plate 13).

I took some photographs of the preparation, one of which eventually found a permanent place on my study wall: the first unequivocal visual record of a cell in which the genes of man and mouse had been amalgamated into a single unit. It would have been enough to sit back and admire it; but if those fused cells were to be anything more than fodder for journalists, there were some crucial questions to be answered. The information that Okada had produced about virus-induced cell fusion had been seminal. However, although he had been working on this problem for more than six years, he had not published anything about the subsequent fate of the fused cells. It was not clear whether, or for how long, they survived, or what normal cellular functions they were able to carry out. Since, until then, the cells that Okada had fused together had come from the

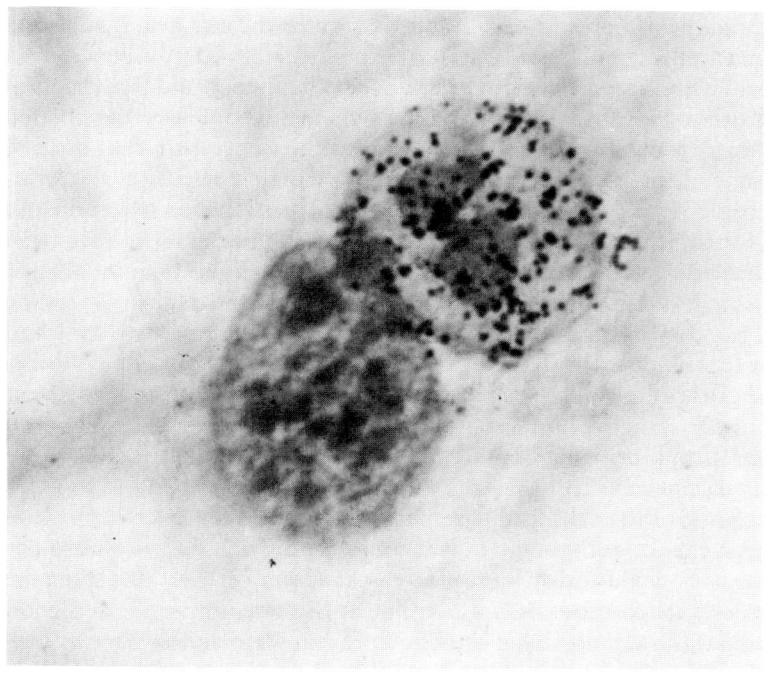

Plate 13. A mouse cell and a human cell amalgamated into a single unit. The human nucleus has been radioactively labelled and shows black grains in the overlying film. The mouse nucleus is not labelled.

one species, it was reasonable to expect that they would have been faced with fewer problems than heterokaryons made with cells from different species. The absence of any information from Okada about the fate of his fused cells was therefore ominous.

I immediately set up experiments to see whether the heterokaryons we had made could carry out the cardinal processes of all living forms: the synthesis of DNA, RNA, and protein. This I also did by autoradiographic techniques and was soon able to show that within the heterokaryon both the human and the mouse nuclei were functional. Both sets of genes were transcribed into RNA and both were in due course replicated. Nor did the synthesis of protein in the heterokaryon seem to be any different from that in the parent cells from which it was derived.

I initially had some difficulty in deciding whether the fused cells could divide or not. As I continued to watch the cultures, it appeared that cells containing several nuclei did not usually succeed in achieving complete cell division, although they did attempt the process. However, cells with only two or three nuclei often divided into two daughter cells, and I formed the impression that at least some of these daughter cells had only a single nucleus. In the autoradiographs I saw some large cells that were labelled over one half of the nucleus only, and I thought that such nuclei must have been formed by the fusion of a labelled HeLa nucleus with an unlabelled Ehrlich nucleus. It later turned out that the mechanism by which these composite nuclei were formed was more complex than I had at first assumed, but the fact that such nuclei were present at all was of great importance for it demonstrated that we had succeeded not only in bringing human and mouse nuclei together in a single cell but also in bringing human and mouse chromosomes together in a single nucleus.

Some of the cells in the preparations stained for microscopy had been caught at the stage of cell division in which the chromosomes are packaged for distribution to the daughter cells. At this stage the individual chromosomes are visible as discrete units and, in a good preparation, their shape can often tell you what animal species they are from. Human and mouse chromosomes are easily distinguishable even to an untutored eye, and although I was not then, and still am not, much good at chromosome diagnosis, it was obvious to me that some of the dividing cells had both human and mouse chromosomes. I asked Charles Ford, one of the world's most distinguished contributors to the study of chromosomes, to examine our fused cells in more detail, and he was soon able to confirm that some of the cells dividing in our cultures did indeed have a mixed chromosome constitution, part human part mouse.

I was very conscious of the fact that something remarkable had happened. There, before my eyes, were human and mouse nuclei functioning amicably together within a composite cytoplasm, and, in some cells, human and mouse chromosomes functioning amicably together within a composite nucleus. For the first time in evolutionary history, the two sets of genes had been brought together within a common organization and, at least in the short run, they were compatible. I was right about the cells of mice and men: in evolutionary terms, their differences were trivial. I kept thinking of

Wöhler and derived satisfaction from the fact that our experiments too had demolished a restrictive conceptual barrier. I knew that when they were published they would create a sensation.

It was not, however, immediately clear just how useful they would be in providing a practical framework for the genetical analysis of somatic cells. I wasn't sure exactly how the two sets of chromosomes were brought together into a single nucleus, what the long-term fate of cells with such composite nuclei was, or how effective the Sendai virus would be over a wider range of somatic cells, especially the different kinds of highly specialized cells that were found in the animal body. It was obvious that it would take some time to obtain answers to these questions, but I decided, despite the loose ends, to publish what we had already done. It was, in my view, too important to be delayed, and I had done enough science to know that no scientific paper is ever complete. In the last days of 1964 we sent off to *Nature* an article entitled 'Hybrid cells derived from mouse and man: artificial heterokaryons of mammalian cells from different species'.

While we were doing this work, I learnt for the first time of an important discovery made four years previously in Georges Barski's laboratory at the Institut Gustave Roussy in Villejuif. I had been sent for review a volume entitled *Cytogenetics of Cells in Culture*, and in it there was an article by Barski and another by Boris Ephrussi and his colleagues. These two articles dealt with the analysis of a phenomenon first observed by Barski, Serge Sorieul, and Françoise Cornefert in 1960. In a mixed culture containing two closely related cell-lines from the same inbred strain of mouse, they had found, and isolated, hybrid cells in which the chromosome complements of the two parent cells had apparently been combined. Boris Ephrussi, a distinguished geneticist of great originality, appears to have been one of the very few to appreciate the importance of Barski's discovery. Initially together with Serge Sorieul, and later with other collaborators, he had extended the observation to several other mouse cell-lines, and he had made a careful analysis of the chromosome constitutions of the hybrids that arose in the mixed cell cultures.

This work was not at the time well known, except perhaps to those especially interested in chromosome analysis. It was not surprising that I had not come across it before, for Barski's papers had appeared in journals that were not taken either by the John Innes or the Dunn School library, and much of the work discussed by Ephrussi in his article was still unpublished or had appeared in reports of meetings I had never heard of. I was not, in any case, especially impressed by what I read; nor was John Watkins when I showed him the volume. All the work described in the articles by Barski and Ephrussi dealt with hybrids formed between one mouse cell and another. Such hybrid cells could not provide a range of markers adequate for serious genetic analysis. Their formation in the mixed cultures was a rare and uncontrolled event, and how it happened was completely obscure. It at once occurred to me that the hybrids could easily have been produced by contaminant viruses that had by chance induced fusion between the two cell-types in the mixed culture. However, Barski had published a paper which appeared to show that they were produced by the transfer of a nucleus from one cell to another. I later learnt that the pessimistic view I took of the experimental usefulness of these spontaneous mouse-cell hybrids was shared by others. Pontecorvo, who knew about the work, had not thought it worth taking up. It turned out that a great deal had to be done before the precise relationship between our heterokaryons and Barksi's hybrid cells became clear, but concerning the ultimate importance of Barski's discovery, time was to show that Boris Ephrussi was dead right.

The year 1965 was my *annus mirabilis*. On 2nd February I gave my first lecture on cell fusion at one of the research seminars held regularly at the Dunn School in term. Florey had got wind that something was afoot and unexpectedly turned up. He listened attentively but did not join in the excited discussion that followed. As he left, I accompanied him to the front door, and then he simply asked, 'Have you got a couple of chaps to help you with that, Harris?' I don't think he quite understood everything that had been said, but he saw at once that it was important, and he wanted to be sure I had everything I needed to develop the work. I reassured him. Our paper appeared in *Nature* on 13th February and, as I expected, was immediately taken up by the press.

A short note appeared in *The Times* on the same day and two days later there was a cartoon in the London *Daily Mirror*. It showed a collection of chimaeric creatures, part human part animal, travelling together in a compartment of the London Underground. One juvenile man–mouse was asking his father: 'Who was Walt Disney, Dad?' (Plate 14). The following week there was an article in the *New York Times* with the heading 'Two British scientists unite cells of mice

CELL FUSION

Plate 14. The reaction of the London *Daily Mirror* to the news that human and mouse cells had been fused together into single units.

and men into hybrids'. Over the next few weeks I received a torrent of newspaper cuttings from all over the world, some of them reaching an almost hysterical pitch of excitement. From the *Coventry Evening Telegraph* came 'Man–mouse cells may yield secret of life'; from *Welt im Wandel* 'Zell-Bastarde unter dem Microskop'; from *Le Soir* in Brussels 'La foire aux gènes'. This hoo-ha did not rapidly subside. Two years later, *Tit Bits* featured a two-page article entitled 'Why scientists create monsters'. It carried an ominous subheading: 'Fused together . . . a man and a mouse. And the next step could be tree–men'.

The reaction of the popular press is worth a moment's thought. It was clear that our demolition of the species barrier had touched off some pretty deeply held fears. I think they were the same fears as had in the past generated that rich folklore in which humans were wholly or partly transformed into animals by magic spells. Our amalgamation of cells from different species was consistently extrapolated by creative journalists to whole animals, and what they wrote was often accompanied by illustrations of chimaeras little different from the grotesqueries that one found in the paintings and stone-carvings of the Middle Ages. In many cases, of course, it was deliberate sensationalism, but, more often than not, the unease was genuine. John Watkins and I were the magicians, and we were a threat. The BBC did a programme about our work which, in the unfettered hands of an imaginative producer exercising editorial freedom, turned out to be a horror story in which I featured as a latter-day Dr Moreau. Only the best of newspapers discussed the scientific implications of our work, but even they were a little unsure of where it might end up. The great spiritual barrier between man and the rest of the animal kingdom was deeply cherished, and it was not to be breached with impunity.

My reaction to all this excited public discussion of our work was simply to turn my back on it and get on with the job. It was obvious that the first thing I had to do was to explore further the range over which inactivated Sendai virus was effective, and particularly whether it would work with specialized cells taken directly from the body. This was a question of overriding importance for two reasons. The first was that we did not then have, and still do not have, any clear idea of what induces the development of such specialized cells. The process is called differentiation, and if heterokaryons could be made by fusing together different sorts of specialized cells, this might open up a new approach to the problem. The second reason was that both HeLa and Ehrlich cells were highly abnormal: they were long-established cell-lines derived from cancers and they had grossly aberrant chromosome constitutions. This was also true of all the pairs of mouse cells from which Barski and Ephrussi had obtained spontaneous hybrids. It was essential to know whether heterokaryons could be made from cells that had normal chromosome constitutions and carried out normal differentiated functions. I already had two such cells under control, the fibroblast and the macrophage, and I could obtain suspensions of a third, the lymphocyte, from Jim Gowans. To these I added one more—red cells from the blood of chickens.

This might seem an outrageously eccentric choice but there were good reasons for it. In man and other mammals, the red blood cells do not contain any nuclei. As the cells develop from less specialized precursors, they eliminate their nuclei and become reticulocytes, as I have described. However, in birds, fish, and some other orders of animal, the nuclei are retained in the mature red cells, but they are screwed down into a state of complete inactivity. In pathology books such nuclei are described as pyknotic (a fanciful misapplication of the Greek word for thick or dense) and the doctrine is that nuclei in this state are well on the way to dissolution. Among cells that still retained their nuclei, the nucleated red cell presented the most extreme form of specialization that could be found. If I could make heterokaryons with chicken red cells, I could make heterokaryons with anything.

It was not long before I knew that I could. I fused the specialized cells together in different combinations and obtained viable heterokaryons in all cases. Their behaviour was a revelation. Whatever the nucleus might have been doing in its original parent cell, it immediately began to dance to the tune played by the new cytoplasm in which it found itself. Nuclei that did not replicate their DNA began to do so; nuclei that did not make RNA began to do so. Whenever a cell that was active in one or other of these central biological functions was fused with an inactive cell, the active partner imposed its pattern on the inactive one. Fundamental new rules governing the synthesis of RNA and DNA emerged. In the case of the red cell nucleus, the results were spectacular. The shrunken structure that had always been assumed to be undergoing degradation woke up, expanded, and soon became a fully functional nucleus again, something that had never before been observed. It was clear that with inactivated Sendai virus you could fuse any animal cell with any other, across the species barrier and across the specializations imposed by differentiation. For cell biologists Pandora's box had been opened. Spring had not yet come when I submitted a second paper to *Nature* telling the world about 'The behaviour of differentiated nuclei in heterokaryons of animal cells from different species'; and by the summer a third had appeared reporting the results of Charles Ford's analysis of the chromosomes in dividing man–mouse hybrid cells.

Shortly after the initial account of our work had made its way into the newspapers, I received a telephone call from Peter Medawar. It appeared that Boris Ephrussi, who was then working in Cleveland, Ohio had rung Peter to see whether it might be possible to obtain a manuscript of our work, which had come to Boris's notice via the report in the *New York Times*. At Peter's suggestion I sent Boris a copy of the manuscript and a few days later received a congratulatory letter from him. This initiated a correspondence that lasted for some three years. It is possible that these letters may have a little historical interest in that they record what was going on in our heads as opposed to the highly distilled products that we published. Perhaps I shall one day deposit them in some archive where they might enliven the day of a future historian of science with time on his hands.

Boris wrote that he had, of course, thought of using viruses to facilitate the formation of cell hybrids but had not done so because, as far as he was aware, multinucleate cells induced by viruses did not multiply. I informed him that at least some of the fused cells that we had made from human and mouse parent cells did multiply and that Charles Ford was examining the chromosomes in them. None the less, Boris remained sceptical about this for a long time and, in one of his letters, provided a long list of arguments for the view that the hybrid cells that multiplied in our cultures might not have been those that had been fused together by the inactivated virus. I, on the other hand, suggested to him that the hybrid cells that arose in his mixed cultures might have arisen from cells fused together by contaminant viruses of one kind or another. Boris replied that he had had his cultures tested and they contained no detectable Sendai virus.

However, as soon as he heard about our work Boris did put rat and mouse cells together in the one culture and, a few weeks later, he obtained and isolated rat–mouse hybrid cells. It is an interesting testimony to the strength of the hold that the doctrine of individual specificity had on the minds of biologists that Boris did not attempt to mix together cells from different species until he was prompted to do so by our experiments. He had been mixing different kinds of mouse cells together for four years, was acutely aware of the paucity of useful genetic markers in such cells, and knew perfectly well that these would be provided by cells from different animal species. Moreover, the experiment in which he mixed rat and mouse cells together was in no way different from the technical point of view from the many others he had made with different mouse cell-lines. He could have mixed rat and mouse cells together four years previously, but he didn't. Some years later, Pontecorvo, in reviewing a little book I wrote about cell fusion, gave it as his view that it could not have been predicted that cells from different animal species would accept each other when fused together. Of course it couldn't have been predicted, but that does not explain why it was never tried.

For two or three years after our initial burst of activity, I concentrated on the analysis of heterokaryons and did not investigate the long-term fate of the mononucleate hybrid cells to which they gave rise. I did, however, regard it as essential, especially in the light of Boris Ephrussi's criticisms and George Barski's claim that his hybrids arose from the transfer of nuclei between cells, to clarify exactly how mononucleate hybrid cells were formed. I did this by analysing cinematographic records of the behaviour of individual heterokaryons monitored continuously over long periods of time. It turned out that my initial assumption that the composite nuclei were formed by direct fusion of the parent nuclei in the heterokaryon was incorrect. What actually happened was that when the heterokaryon began to divide and the chromosomes were being packaged for distribution to the daughter cells, all the chromosomes from the two parent nuclei were brought together into a single linear array; and when the cell actually divided, this linear array split down the middle and provided each daughter cell with two complete sets of chromosomes, one from each parent cell. These two sets of chromosomes were then collected into a single nucleus and got to work again.

The mononucleate hybrid daughter cells were themselves capable of further division and generated small colonies. We had noticed these colonies, obviously quite different in appearance and growth character from the two parent cells, quite early in our studies, but the definitive proof that they were colonies of man–mouse hybrid cells had to wait until we had produced appropriate immunological reagents and devised methods that permitted us to demonstrate human and mouse proteins within the one cell.

We were not able to isolate populations of these hybrids as Boris Ephrussi had done with his rat–mouse hybrids. This was because the man–mouse hybrids grew very slowly and were inevitably overgrown by unfused cells. It is still sometimes difficult to isolate hybrids in the presence of more rapidly growing unfused cells, but for certain cell combinations the problem was effectively solved by John Littlefield. He had shown that if pairs of cells were first selected for their resistance to different kinds of toxic drugs, it was possible to devise a medium in which the parent cells could not grow but the hybrid cells could. Boris had used such pre-selected cells and a variant of Littlefield's technique to isolate his rat–mouse hybrids; but the cells we were working with had not been selected in any way for resistance to drugs, and Littlefield's technique was inapplicable. Boris's opposition to the idea that cells fused together by viruses could multiply was finally disposed of by George Yerganian and Marilyn Nell, of the Children's Cancer Research Foundation in Boston, who, about a year after our initial paper had appeared, used inactivated Sendai virus to fuse together cells from two different species of hamster and succeeded in isolating proliferating populations of hybrid cells from this cross.

My life changed radically. From having been for years an outsider fighting for an eccentric point of view, I suddenly found myself with a front seat on a world-wide band wagon. Invitations to lecture came streaming in from all over the world, and I was still young and foolish enough to be flattered. Many years later a curious set of circumstances induced me to take an interest not only in the work, but also in the life, of Otto Warburg. It appears that he reserved his greatest contempt (with which he was pretty liberal) for scientists who spent their time travelling around and lecturing about their work or, more often, the work of their junior colleagues. He called them 'itinerant preachers' (Wanderprediger) and regarded them as a blight on the scientific landscape. I'm afraid that, for a while, I came pretty close to being an itinerant preacher.

There were, however, two trips that I made in 1965 that were more to me than mere scientific junkets. In April, at Hämmerling's invitation, I visited his institute in Wilhelmshaven. Wilhelmshaven had been almost totally destroyed during the War and had been replaced by a nondescript new town of no discernible character. However, out on the flat dunes behind a sea-wall, an old rectangular Wilhelminian building had survived, and there, in the chaos of postwar Germany, Hämmerling had started his experiments again. Two or three times I took a walk with him along the dyke, and we talked about the War, the period that had preceded it, and about science. He was a sad and disillusioned man, half-resigned to the fact that his work was grotesquely underestimated; but when he spoke about *Acetabularia* the old fire was still aglow.

In the summer I visited Brno and Prague to take part in a memorial symposium to mark the centenary of Mendel's great work on the hybridization of plants, the foundation stone of the science of genetics. There I met Boris Ephrussi for the first time and his altogether charming young daughter Anne. Needless to say, there were matters of common interest on which Boris and I didn't see eye to eye. As it happened, I spent most of my time in Brno in the company of Jo Hin Tjio, who had recently shown that human beings had forty-six chromosomes instead of forty-eight as had until then been supposed. He took a photograph of the little garden in which Mendel had done his classical experiments and later sent me a print of it. This, in due course, also found a place on the wall of my study; and on my desk there is a small transparent box containing soil from that same garden.

I got an early indication of the speed with which virus-induced cell fusion was being taken up in other laboratories by the arrival of requests for Sendai virus. This began as a trickle shortly after our first paper appeared, but within a year or so the demand had reached such proportions that I had to set up a regular supply-line within the department to grow and test the virus not only for our own needs but for dispatch to countless other destinations around the world. The requests for virus were followed in a surprisingly short time by a surge of publications in which cell fusion was used for one purpose or another; and that flow has continued unabated to the present day.

For me the problem at the end of 1965 was to decide which of the limitless experimental possibilities opened up by cell fusion I was going to concentrate on. Although it was the idea of applying genetical methods to somatic cells that had first sparked my interest in cell fusion, I did not at once begin genetical experiments with hybrid cells. The sight of the red cell nucleus rising, like Lazarus, from the dead had captured my imagination and I decided I'd start with that. Cell fusion had moved nuclear RNA from the centre of my attention, but I was naturally still intensely interested in the flow of genetic information from nucleus to cytoplasm and, of course, in the flow of signals from cytoplasm to nucleus. The reactivation of the red cell nucleus seemed to present an entirely novel way of exploring these questions.

My belief that cells from different animal species would accept each other when fused together was based on the assumption that the fundamental mechanisms that ordered events within the cell would be very similar throughout the animal kingdom and hence compatible. However, a hybrid cell was formed by the amalgamation of the two parental cytoplasms as well as the parental nuclei, so the possibility existed that each set of genes might be responding only to signals from its own set of cytoplasmic components. A form of segregation could be envisaged in which the two systems of signals did not interact, the mouse genes responding only to mouse cytoplasmic signals and human genes only to human cytoplasmic signals. If the hybrid cells had indeed been produced by the transfer of a nucleus from one cell to another, as Georges Barski had proposed, then this possibility could hardly have been entertained, for, at least initially, the transferred nucleus, lacking its own cytoplasm, would have had to be responding to signals emanating from the foreign cytoplasm into which it had been deposited. But I now knew that hybrids were produced by cell fusion, not by nuclear transfer, and the question remained open. Heterokaryons made with chick red cell nuclei provided the answer.

Eveline Schneeberger, who was spending a sabbatical year with me on leave from Harvard, had been examining the formation of those heterokaryons with the electron microscope and had found that, before cell fusion occurred, the membranes of the chick red cells were perforated by the action of the virus, so that their cytoplasmic contents leaked out. In this case, therefore, the chick red cell nucleus was deposited into a foreign cytoplasm unaccompanied by its own cytoplasm. When it was reactivated it must have been responding to signals that emanated from the foreign cytoplasm, human or mouse as the case may be. But how well did it respond?

This I explored with a series of talented D.Phil. students, Eric Sidebottom, Peter Cook, and Ilan Déak, and we were soon able to show that a normal flow of RNA was established from the reactivated red cell nucleus to its new cytoplasm and that, in due course, proteins coded by genes in that nucleus were synthesized in the hybrid cell. There was thus no doubt that the chick nucleus responded to the instructions given to it by the human or mouse cytoplasm and understood those instructions perfectly well. In the work on red cell nuclei I was joined by Nils Ringertz, then a member of staff, and later Director, of the Institute for Medical Cell Research and Genetics in Stockholm. Nils made the study of reactivated red cell nuclei a speciality of his laboratory and continued to make important discoveries about their behaviour long after the centre of my own interest had drifted elsewhere.

Heterokaryons made with red cell nuclei were not, however, the only fused cells that were then being studied intensively in the laboratory. We had still to discover how cells in widely different physiological states and at different stages of progress through the cycle of cell division managed to get themselves synchronized when they were fused together. This was sorted out by Bob Johnson, another D.Phil. student, who in the process made an observation that later provided a standard technique for the study of such problems. Yet another D.Phil. student, Fiona Watt, finally explained how it was that all the chromosomes in a cell with two nuclei were brought together into a single linear array when the cell divided. During the course of my scientific life I have had some thirty D.Phil. students and I should like at this point to salute them. In the matter of intellectual exchange, there is no question but that I got as much from them as they from me. They are now scattered all over the world, and distance precludes my seeing many of them as often as I should like; but I can't think of a single one, temperamental or placid, punctilious or cavalier, with whom I would not gladly work again.

I did eventually get round to doing genetical experiments with hybrid cells. It turned out that hybrids between human and mouse cells were by chance more useful than any other combination for one of the classical applications of genetical methods: determining where in the set of chromosomes particular genes were located. Studies on proliferating hybrid cells had shown that, with time, a reduction took place in the number of chromosomes that they contained. By a mechanism that is still not absolutely clear, some chromosomes were progressively eliminated from the cell. In 1967 Mary Weiss and Howard Green, of the New York University School of Medicine, noticed that in a line of man–mouse hybrid cells that they had constructed, the human chromosomes were rapidly and preferentially eliminated, leaving cells with a more or less intact set of mouse chromosomes but with incomplete and variable human chromosome sets. It was at once obvious that in such cells one could see whether the loss of a human genetic marker was regularly associated with the loss of a particular human chromosome. If so, one could infer with confidence that the gene coding for that marker was located on that chromosome.

This procedure was rapidly adopted by many laboratories and within a few years resulted in two or three hundred human genes being assigned to specific human chromosomes. However, compared with what was possible with bacteria or even with that classical test object of animal genetics, the fruit fly *Drosophila*, the mapping of human genes by chromosome elimination was primitive. At a meeting that I attended in Paris, François Jacob, no doubt having in mind the precision of his own work with *Escherichia coli*, expressed the view that genetics could hardly be said to have begun until one could establish the linear order of genes in the chromosome and determine the distances between them. Mapping the location of genes by chromosome elimination couldn't, of course, do that.

I took François Jacob's challenge back to Oxford and began to worry about it. What I came up with was based on an old observation concerning the effects of X-rays on cells. It was known that X-rays induced breaks in the chromosomes and that the number of breaks was proportional to the dose of X-rays given. It seemed obvious that if two genes were far apart, then the chance that they would be segregated by a break in the chromosome would be greater than if the two genes were close together. Now it seemed to me that if the human parent cell were irradiated with an appropriate dose of X-rays before being fused with the mouse cell, and if the broken human chromosomes were then eliminated preferentially from the hybrid cell, then genes located in close proximity to one another would tend to be eliminated together, whereas genes that were far apart would more frequently be eliminated separately. I thought it might be possible to extract from measurements of the frequency with which genes were eliminated together or separately some estimate of the distances between them and also possibly information about their linear order. I put this proposition to a new D.Phil. student who had just joined me, Stephen Goss, and he developed it with a virtuosity that I have rarely seen in a novice to research. An accurate and reliable method for estimating the distances between genes and for determining their linear order emerged. It is still occasionally used, although the whole strategy of gene mapping has since been transformed by the application of new advances in the methodology of handling DNA.

Cell fusion brought to the Dunn School a continuing stream of overseas visitors, from pre-doctoral fellows to departmental heads. One of the first was Earl Benditt, Chairman of the Department of Pathology at the University of Washington in Seattle. Earl became a regular visitor to the Dunn School and two of his staff also spent sabbatical years with us. Indeed, a sort of *concordia amicabilis* developed between the two Departments, and it was as Walker-Ames Visiting Professor in Seattle that I was initiated into the pleasure of sailing on Puget Sound and the glory of West Coast cracked crab.

Earl was succeeded by a cohort of distinguished bacterial geneticists led by Ed Adelberg and Ellis Engelsberg. Bacterial geneticists at once saw the possibilities opened up by cell fusion, and many of them moved quickly to apply their highly refined skills to somatic cells. The bacterial geneticists were followed by human geneticists: in order of appearance, Jack Miller, Harold Klinger, Eric Engel, Bill Kelley, Ed Seegmiller, David Martin, Ernest Chu, and John Littlefield. And as the use of cell fusion spread into disciplines other than genetics, biologists from the most unexpected quarters turned up. I began to feel very satisfied with the way pathology at the Dunn School was going.

Americans always predominated. Just how many of them had worked at the Dunn School over the years was brought home to me one evening at Harvard where, in the autumn of 1969, I was giving the Dunham Lectures. A small group of people, all of whom had at one time or another taken sabbatical leave at the Dunn School, were busy buying me cocktails, when an outsider came into the room. As soon as he saw us, he threw up his hands and exclaimed: 'Aha! The Dunn School Mafia'. It wasn't an altogether facetious remark, for there are few major scientific centres in the USA where we don't have at least one representative (Plates 15 and 16).

Although Oxford is, in my view, obsessed with undergraduate teaching, it offers little in the way of systematic post-graduate instruction. D.Phil. students in science subjects simply get down to work at the bench, and they are left to their own devices to acquire familiarity with any technique or background knowledge that might be necessary for their research. There is no formal course-work and no requirement to attend any lectures or seminars that might be given. I think this is a much better system for intelligent graduates than one in which a year or more of their time is first taken up with compulsory attendance at what are simply upgraded undergraduate courses in subjects thought to be of special relevance to their work. However, the Oxford doctrine of *laissez-faire* does tend to break down in fields where developments are moving so quickly that they are not yet represented in the undergraduate curriculum. When I came to Oxford, cell biology as a coherent discipline did not appear anywhere in the syllabus of undergraduate teaching, and very few of the college dons could cope with the subject at an advanced level.

This deficiency and, more deeply felt, a need to put my own view of the cell to an intelligent audience induced me to offer a set of

Plate 15. Some prominent members of the Dunn School mafia in the USA. (a) Robert H. Ebert, sometime Dean of the Harvard Medical School. (b) Robert Q. Marston, sometime Director of the National Institutes of Health. (c) Earl P. Benditt, sometime Chairman of the Department of Pathology in the University of Washington at Seattle. (d) Philip P. Cohen, sometime Chairman of the Department of Biochemistry in the University of Wisconsin at Madison. All of the photographs were taken during the periods that each of these men spent working at the Dunn School.

Plate 16. Some prominent members of the Dunn School mafia in the USA. (a) William R. Barclay, Chief Executive Officer of the American Medical Association. (b) George B. Mackaness, President of the Squibb Institute for Medical Research at Princeton. (c) Vincent T. Marchesi, Chairman of the Department of Pathology at Yale University. (d) John W. Littlefield, Chairman of the Department of Paediatrics at Johns Hopkins Medical School, Baltimore. All of the photographs were taken during the periods that each of these men spent working at the Dunn School.

special lectures under the general title 'Nucleus and cytoplasm'. The response was astonishing. Even the stairs in the Dunn School lecture theatre were crowded out, and the audience stayed with me to the end. Afterwards, I was urged repeatedly to publish the lectures, and that is how *Nucleus and Cytoplasm*, my first book, was born. The three editions it went through over the next five years record the changes that took place in my point of view as increasing knowledge proved me right in some respects and wrong in others, much as you would expect. Most scientific books these days are compilations of chapters written by many different hands, and I am glad that I made the attempt to present one man's view of a large subject, however imperfect it might have been. The Clarendon Press at Oxford, which published the book, kept asking for a fourth edition, but by that time cell biology had changed so much that I felt that, if I were to write another book on the subject, it would have to be a very different book. I haven't yet written it and perhaps never shall.

From time to time Florey asked me to Queen's for lunch or dinner. With age he had become almost unrecognizably mellow although his conversation never quite lost its characteristic touches of astringent scepticism. I didn't often see Lady Florey for she had become increasingly incapacitated by illness and was not much in evidence in the Provost's Lodgings. She died at the beginning of October in 1966, and a few months later, at the Old Register Office in St Giles, in the presence of two witnesses and no one else, Florey married Margaret Jennings who was then still teaching at the Dunn School. On the morning of 22nd February 1968, Jim Kent came in to tell me that Florey had died during the night. In accordance with his wishes he was cremated, but a conventional funeral service was held in Old Marston at the Church of St Nicholas, which was just over the road from his house but which, during his lifetime, he had never frequented.

10
CANCER

If you do something dramatic in science, certain ceremonial acknowledgements of your achievement inevitably come your way. These take the form of prizes, medals, honorary degrees, memberships of academies, and so on. Most scientific biographies seem to me to be marred by an excessive preoccupation with all this froth. I can hardly believe that the average reader could be enthralled by it. I don't therefore propose to burden these pages with accounts of my own harvest. There's a fair sampling of it in *Who's Who*, and anyone misguided enough to be interested can look it up there. There were, however, two non-scientific consequences of my work on cell fusion that are perhaps worth more than a casual mention. The first was that I was offered a number of jobs, all of them more lucrative than being a professor at Oxford, and it is just conceivable that my reasons for turning them down might be of interest. The second was that I became increasingly involved in the administrative apparatus that implements public support of science in the United Kingdom. I remained involved for long enough to form strong opinions about this activity, and I had no difficulty when the time came in deciding that I did not want to make a career of it.

The 1960s and early 1970s were an affluent, expansionist era for science. Old institutions were everywhere being enlarged and new ones were springing up all over Europe and America. Hardly a year went by without my receiving two or three enquiries about whether I might be interested in some position or other. I shall leave it to the unlucky individual who will be given the task of composing a biographical memoir about me for the Royal Society to compile the full list from my records. But I'd like to say something about a few of these jobs because they made me pause to think. The first was the offer of a chair at Harvard. This was made while I was giving the Dunham Lectures there. In fact, there were two proposals, one from the Biological Laboratories and one from the Medical School, but when I declined the first, and my reasons became known, the second was not offered. What impressed me most was the speed with which a decision of that kind could be taken and the enterprising way in which the offer was made. It was absolutely unthinkable that the University of Oxford could have made such a spontaneous gesture even if the lecturer had been Isaac Newton. And although we did not get down to details, the offer was in all practical respects immensely attractive.

Of all the centres that I had visited in the USA during my stay there and subsequently, Harvard was the place where I felt most at home. Why did I not accept? For sentimental reasons. I had begun to acquire a primitive sense of loyalty to Oxford. It had given me a job when I needed one, and then given me the best job it had in my line. Although the scale on which I operated was modest by American standards, it was big enough for me, and, by courtesy of the Cancer Research Campaign, I had what I required to get on with my work. I do not make any criticism of those who, in circumstances less fortunate than mine, decided otherwise; but I found it difficult to cross the Atlantic simply because the pastures there were greener. Besides, for me, already an expatriate, there was a deeper reason. I had torn up one set of roots and was only very slowly growing another. I didn't want to tear them up again. The people at Harvard were a decent lot, and they did not press me once I had explained why it was that I was not on the market. However, when I look now at the problems facing the universities of the United Kingdom I sometimes wonder whether I made the right decision.

In 1971 ill health obliged Peter Medawar to resign the Directorship of the National Institute for Medical Research. John Gray, who had succeeded Harold Himsworth as Secretary of the Medical Research Council, came down to Oxford to ask whether I might be interested in the position. This was very flattering, for in the eyes of officialdom, the Directorship of the National Institute was the top job in British medical research. I promised that I would give the proposition some serious thought. And I did; but in the end I came to the conclusion that it was not for me. The fundamental reason was that I did not want to stop being a scientist and become a scientific manager. There's nothing wrong with being a scientific manager, but it's much less exciting than being a scientist. I had found that, with a little self-discipline and a little energy, it was altogether possible to be the Professor of Pathology at Oxford and still remain deeply involved in experimental work. I very much doubted whether this would be possible for the Director of the National Institute of Medical Research.

The Institute was perhaps ten times the size of the Dunn School, unavoidably organized and administered in a rather formal way; and although it was pocket-sized compared to the National Institutes of Health in Bethesda, it was still, in my view, far too big for comfort. Peter Medawar had once told me that he had arranged matters so that he could spend two whole days each week in the laboratory. That might well have been enough for him and the type of work he did, but it certainly wouldn't have been enough for me. What's more, my experience at the John Innes of the inconstancy of government research policy made me uneasy about committing my future once again into bureaucratic hands that did not themselves have complete control of the situation. I could foresee a time when ministerial decisions might make life at the National Institute of Medical Research, or at any other research institute funded directly from government sources, very unpleasant. Finally, there were practical considerations that made the move from Oxford to London unattractive on the revised terms that applied to the Directorship after Peter's retirement. John Gray seemed genuinely sorry that I didn't accept the offer; and I couldn't help feeling a little guilty that I had not been able to make the public-spirited gesture that appeared to be expected of me.

One morning, out of the blue, a letter arrived from Canberra. It came from Dr H. C. Coombs, the Chancellor of the Australian National University, and it contained the astonishing suggestion that I might be interested in the Vice-Chancellorship. Would I come out to Canberra and have a talk about it? There was obviously a conspiracy afoot to put me out of business as an experimentalist. I wrote back to say that I was very moved, which indeed I was, but that I was not yet ready to down tools and become a full-time administrator. However, it wasn't long before another letter came from Canberra, this time from the man who did accept the Vice-Chancellorship, asking me whether I would be interested in the Directorship of the John Curtin School of Medical Research. This too, I thought, would be essentially a managerial job, but of course it had special associations for me.

It was not easy to forget the circumstances which had finally induced Florey to turn the job down and which, at the same time, had squashed my interest in the John Curtin School. Unless I was very wrong in the impression I had of the way the Australian National University worked, it seemed unlikely that this directorship would be an attractive position for me. However, I thought that this might well be the last chance I would get of exploring the possibility of returning to Australia, and I could not turn the offer down out of hand. I therefore wrote to the Vice-Chancellor and explained, once again, that I could not make an informed decision about his offer without having first spent some time in Canberra together with my family. Our children had all been born and brought up in Oxford, and they felt perfectly at home there. Unlike mine, their roots had not been fractured, and I doubted whether they would have any desire to transplant themselves to an environment with which they were familiar only by hearsay.

To my astonishment, my letter produced a most generous response. I was welcome to come out to Canberra with my family for as long as I thought necessary, and the Australian National University was very willing to provide the means. I suggested a month during the English summer vacation. In addition, I proposed that I should give a series of lectures while I was in Canberra, so that if, as seemed to me probable, I was not in the end able to accept the Directorship, my hosts would not regard the exercise as a total waste of time and money. It must have been an affluent time in the Australian National University, for I soon received five first-class round-the-world air tickets, which enabled the Harris family in due course to become connoisseurs of the front-ends of Boeing 747s.

Canberra was a visual delight, especially the gentle Monaro sheep pastureland that surrounded it. The melancholy piping of the magpies accompanied my morning shave, the sun shone, the evening air carried its full load of nostalgic messages. I gave my lectures, which seemed to be appreciated by a lively audience; and we were lavishly entertained in many homes. But when it came to the Directorship of the John Curtin School, I found the job enmeshed in bureaucratic complexity that left little room for initiative and little time for anything but politics and administration. Florey had once said that the administrators in Canberra did not realize that the position was not one that a working scientist would find particularly attractive. He was right. The moment of final decision came for me when I was asked to meet an academic board of some kind whose approval was apparently necessary for the appointment to be made. Some members of this body behaved as if they were interviewing an applicant eager for a job. I disillusioned them. The children were delighted by the countryside, the bright birds, the unfamiliar animals; but as the month drew to its close they all thought it was time to go home.

For many years after the War, I declined invitations to visit Germany. My reactions to the recent history of that country, exacerbated by my own origins, were still too strong for me to be able to associate on friendly terms with people whose behaviour during the Hitler period had been equivocal. However, after my visit to Hämmerling's laboratory, my contacts with German scientists increased, and I began to meet young men whose scientific style and whose outlook on the world I found admirable. I gradually came round to the view that, if asked, I had a moral obligation to help those who were striving to establish a humane and liberal society in Germany. If one declines to support the forces for good, one supports the forces of evil. A humane and liberal society cannot exist without humane and liberal science; and I decided that if I could help in the effort to achieve this, I would. So it came about that, with increasing frequency, I found myself being asked to advise on scientific developments in Germany, especially in the Max Planck Society, of which I eventually became a foreign member.

The Max Planck Society is the major funding agency for the support of basic scientific research in the Federal Republic of Germany. It administers some sixty research institutes, and it was in connection with one of these that I received the last of the job offers that I am going to describe. It came in the form of an invitation to visit Berlin to discuss the future of the Institute for Cell Physiology which, until his death the year before, had been directed by Otto Warburg. I was met in Berlin by Feodor Lynen, a distinguished biochemist whose hospitality I had already enjoyed on previous occasions. It transpired that one of the plans envisaged by the Max Planck Society for the future of Warburg's institute had me at its centre. I cannot imagine how the idea could have gained currency that I might consider leaving Oxford for Berlin, but when Lynen began to talk about the future development of cell biology in Germany, I found to my amazement that I was being offered the Directorship of the Institute on terms that I can only describe as *carte blanche*.

Berlin was then a city where it was easy to talk to ghosts: they called out to you as you walked the streets. But I don't think I ever had a more intimate conversation with a ghost than when I visited Warburg's Institute. It had been built specifically for him and was modelled on some East Prussian junker castle of which he was particularly fond. Nothing had been touched since he died. In the basement skilled technicians were still making measurements in accordance with protocols that he had written out eighteen months previously. His library and the desk at which he used to sit were undisturbed. His manservant and life-long friend, Jakob Heiss, and his Great Dane, Norman, were still there, and I was introduced to both. I was shown the house in which he had lived, set in a sizeable plot of arable land where he grew his own grain and vegetables in order to escape contamination by modern food additives. An unchallenged autocrat in an East Prussian heaven.

In Oxford a professor exercises power by the consent of his colleagues. If this consent is withdrawn, formal mechanisms can be brought into play to ensure that professorial power is effectively neutralized. I have never felt it necessary to have any greater power than that sanctioned by the confidence of my colleagues, and I do not think that Lynen can have had any idea how unattractive I found the feudal atmosphere of Warburg's castle. It will surprise no one that I did not decide to exchange Oxford for Berlin, despite the extravagant financial and scientific inducements; but I continue to regard it as no small compliment that, in German eyes, I should have been seen as a plausible scientific successor to Otto Warburg.

As the years and the jobs went by, it became clear to me that, unless something totally unforeseen happened, I would end my days in Oxford, a peaceful prospect that I did not view with displeasure. That peace now began to be threatened in another way. There is a widely held belief that the creative powers of scientists wane rapidly in middle age. It is certainly true that the great majority of them begin to distance themselves from the bench and assume an increasing load of public work of an advisory or managerial kind. I am, however, unsure whether it is the loss of creativity that drives scientists in that direction or whether it is the increasing load of public work that stifles the creativity. It is certainly much more difficult to do creative experiments than to give advice, and it is perfectly understandable that with increasing age there should be a drift away from the more exacting and to the less exacting life, especially as the rewards of vigorous participation in scientific administration can be considerable. However, as I look about me, I view with regret the departure from real science of so many gifted individuals whose powers do not show the least sign of failing, but who have been seduced away by a system that seems to me to be specifically designed to put talented scientists out of business.

For me the process of seduction began with an invitation to serve on the Biological Subcommittee of the Science Research Council. This was followed by the Agricultural Research Council, the Council of the Royal Society, the Scientific Committee of the Cancer Research Campaign, the Governing Body of the Lister Institute, the Scientific Advisory Committee of the European Molecular Biology Laboratory, and so on and so on. To begin with, each new undertaking had its interest, but in my case that interest was never very intense, and although I of course acknowledged that I had a duty to pull my weight, I was always very conscious of the progressive erosion of my time.

My term of service on the Agricultural Research Council and the Council of the Royal Society coincided with the deliberations that were taking place in government circles as a consequence of Victor Rothschild's report on the organization of civil science. Rothschild's main theme was that science should be reorganized on an essentially commercial basis. The appropriate government departments would operate as customers, who would commission the work they wanted done, and the scientists would operate as contractors, who would carry it out. The trouble with this was that the government departments turned out to have no idea what they wanted done, and the scientists, at least the good ones for whose services there is always an eager world-wide market, simply went elsewhere. It seems to be very difficult for politicians and civil servants to take it on board that only in times of national peril will creative scientists agree to take instructions about what they are going to do.

My experience of politicians during this period left me with the depressing conclusion that, with very few exceptions, they will need a great deal of education before we can hope to have a rational science policy in the United Kingdom. The other depressing conclusion I reached as a result of all this busy work on committees was that the 'peer-review' system simply didn't work or, at best, worked very inefficiently. The peer-review system is a committee structure that was devised to ensure that every application for the support of research was assessed by a committee composed of scientists who were judged to be the applicant's peers. In practice, only one or two people around the table are at all competent to deal with any particular application in depth, and their views, which may or may not be impartial, dominate the assessment. The other members of the committee usually know little or nothing about the subject in question and vote on the strength of superficial personal impressions. Peer review, as I experienced it, was, for the most part, little more than the unweighted mean of generalized ignorance; and it consumed an immense amount of time. I am sometimes asked what I would put in its place. I do have some ideas about this, but it would take another book to set them down in detail, and I am not at all sure that anyone would want to read it.

In addition to all this committee work there was the unrelenting pressure of invitations to travel, subtle temptations for me to be anywhere in the world but where I ought to be. Lectures, conferences, seminars, workshops, consultations, ceremonies, what have you. One morning, as I got dressed in the winter dark to catch an early train to London, it dawned on me that I didn't really like the way my life was going. It proved to be a particularly futile and irritating day in London, and when I got back home late that evening, I decided that the rot had to stop. I resolved that, at the earliest possible opportunity, I would shed my recurrent London commitments, or at least reduce them to proportions that did not make unacceptable inroads into my experimental work; and I also resolved that I would accept invitations that took me away from Oxford only if there was a very good reason for doing so.

The word soon got around that Henry Harris didn't like travelling and that he invariably refused invitations to take part in conferences or seminars. The flow of invitations gradually abated. It was not long before the only regular duty I had in London was attendance at the meetings of the Scientific Committee of the Cancer Research Campaign, to which I acknowledged a special allegiance. Before I left the Agricultural Research Council, it had become clear that the Rothschild proposals were going to be accepted. My opposition to them had been public and was well-known; but that did not stop my receiving an enquiry to see whether I would consider accepting the Secretaryship of the Agricultural Research Council and, even more incongruous, a similar enquiry about the newly created post of Chief Scientist at the Ministry of Agriculture, Fisheries, and Food. Mysterious are the ways of government.

So what was going on in the laboratory that induced me to give it such an overriding priority in the gamut of my activities? Well, I had finally got round to having a look at the cancer problem. Before revealing what I was up to, I must, however, make a small digression to explain what dominant and recessive genes are. Each cell in the body contains a complete set of chromosomes from the male parent and a complete set from the female parent, so that every gene, except for certain special cases that are irrelevant in this context, is present in the cell nucleus in two copies, one paternal and one maternal. Now it can happen that a paternal gene produces a particular cellular characteristic that differs from that produced by the corresponding maternal gene. In that case, the paternal gene is said to be dominant if the cell shows the paternal, but not the maternal, characteristic; and the maternal gene is then said to be recessive. On the other hand, if the maternal characteristic is expressed in the cell, but not the paternal, the maternal gene is the dominant one and the paternal gene recessive. If one of the two corresponding genes undergoes a mutation that has the potential to change the cellular characteristic determined by that gene, then the mutation is said to be dominant if that cellular characteristic is in fact changed; but if the mutation is overridden or compensated for by the activity of the corresponding unmutated gene, then the mutation is said to be recessive (Fig. 3). When I began my investigations, it was widely held (and still is) that a normal cell becomes cancerous because it sustains mutations that render it insensitive to the normal mechanisms that control the growth of cells in the body. However, no one had any clear idea of how many mutations might be involved, where they were located, or whether they were dominant or recessive.

Because cancer has such an inexorable image, it was generally assumed that the mutations that caused it would have to be dominant. The very few experiments that had so far been done to examine the question supported this assumption. These were experiments done on hybrids that arose spontaneously in mixed cultures of mouse cell-lines. In some of these mixtures one of the parental cell-lines happened to be malignant and the other not. (A cell-line is said to be malignant if, when injected into an appropriate host, it grows progressively and kills the animal.) What had been reported was that the hybrids formed by the fusion of the malignant cells with the non-malignant ones were themselves malignant. It was therefore concluded, quite reasonably, that the mutations determining malignancy were dominant mutations.

Once again I was fated to see things differently. I found it very difficult to believe that cancers could, in general, be caused by

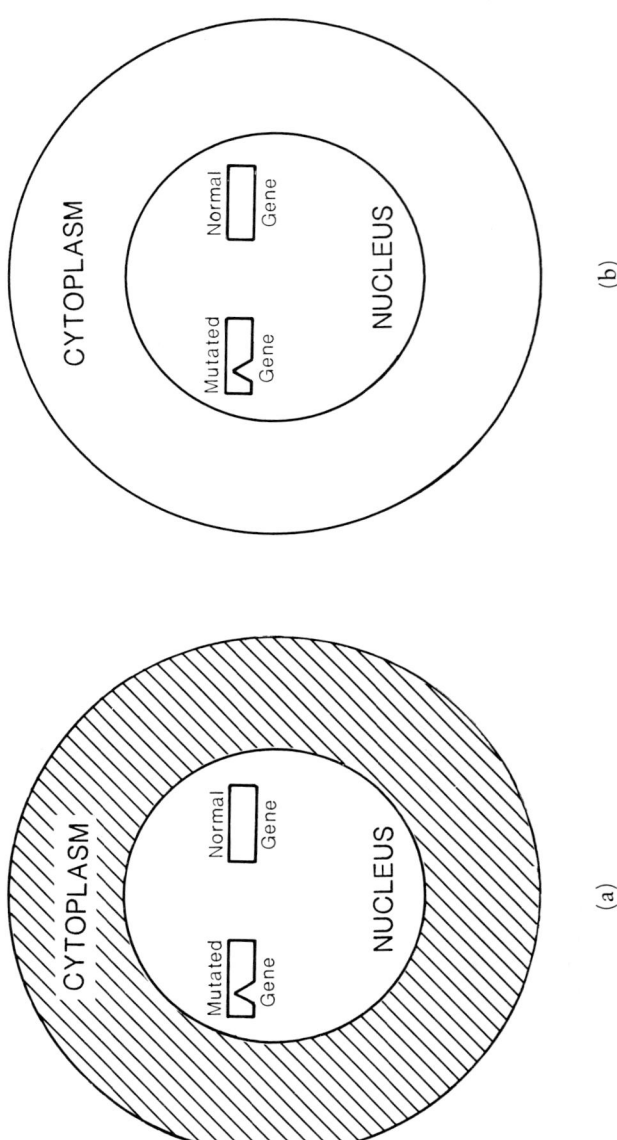

Fig. 3. (a) A *dominant mutation*: one of the two corresponding genes in the cell is mutated, the other is not — if the cell shows an *abnormal* character, the mutation is said to be dominant. (b) A *recessive mutation*: one of the two corresponding genes in the cell is mutated, the other is not — if the cell shows a *normal* character, the mutation is said to be recessive.

dominant mutations. A few simple, perhaps too simple, calculations, based on the frequency with which mutations were known to occur and the number of cells in the body, convinced me that if cancers were caused by dominant mutations, that is, mutations that produce a change in some cellular characteristic despite the presence within the same cell of a corresponding gene that has not been mutated, then none of us would ever have reached adulthood without succumbing to a cancer of one kind or another. Indeed, the cancers would probably already have arisen during the development of the foetus. What was more, I could see technical deficiencies in the experiments that had been done with crosses between malignant and non-malignant cells that led me to doubt the interpretation that had been given them. Since I now had the ability to produce hybrids from any combination of cells I wanted, I decided to examine the question of dominance in a more systematic way. It was, in any case, difficult to envisage a matter of greater importance for our understanding of how cancer cells were generated.

I began the work with Jack Miller who was spending a sabbatical year with me on leave from Columbia University. I fused some malignant cells with non-malignant ones, got out the hybrids, and injected them into appropriate animals to see whether they would grow progressively. They did. So, to that point, our findings agreed with those that had been reported. But when Jack came to look at the chromosomes of the hybrid cells we got our first surprise. The cells that grew progressively in the animal were not identical with the ones we had injected: they all showed substantial losses of chromosomes. I have already mentioned that hybrid cells tend to lose chromosomes as they multiply, and I have described how this loss can be exploited to assign genes to specific locations. In the context of the experiments that Jack and I were doing, however, the loss of chromosomes had a special significance. It raised the possibility that the hybrid cells might be malignant, not because the mutations were dominant, but because the corresponding unmutated genes had been eliminated as a consequence of the chromosome loss. It was conceivable that if we could get hold of the hybrid cells before substantial chromosome losses had occurred, we might find that they were not malignant. If this were so, then it would mean that the corresponding normal genes had the ability to suppress the effect produced by the mutated genes. In other words, the mutations would then be recessive, not dominant.

At this point I turned to George Klein, the Director of the Department of Tumour Biology at the Karolinska Institute in Stockholm. George had the best collection of experimental mouse tumours in Europe and the large stocks of inbred mice that would be required to test the hybrid cells properly. I explained my ideas to him and suggested a collaboration in which I would make the hybrid cells in Oxford and he would test them in Stockholm. He agreed. And that led to the most exciting scientific collaboration I have ever had. As soon as I had isolated the hybrids, they were taken to London Airport and were then collected on arrival at Stockholm by one of George's staff. Collaboration over long distances is usually an ineffective and cumbersome exercise, but this one went without a hitch. It lasted for five years and was accompanied by a continuous exchange of letters in which each of us commented on the results as they came out. (George, by the way, is the world's best correspondent.) This collection of letters will also be deposited in the archives in due course. I know of no other instance in which a long and complicated series of experiments has been expounded in this way by the written comments of the participants.

By the time we had analysed three crosses between malignant and non-malignant cells, it was clear that my guess had been right. If we tested the hybrids as soon as possible after isolating them, before they had undergone substantial chromosome losses, we found that they were unable to grow progressively in the test animals. The cells became malignant only after some elimination of chromosomes had occurred. I therefore argued that some gene or genes located in the chromosome set derived from the non-malignant parent cell must have had the ability to suppress, to override, or to compensate for the genetic defect or defects that enabled the malignant parent cell to grow progressively and kill its host. Malignancy was being suppressed by the activity of normal genes. There was a powerful fail-safe device in the cell which ensured that mutations that might lead to malignancy were held in check. Hybrids between malignant and non-malignant cells did not become malignant until this fail-safe device was eliminated by chromosome loss. I regarded these findings as little short of revolutionary and decided that our experiments, although preliminary, were solid enough to warrant publication. We sent off to *Nature* a paper entitled 'Suppression of malignancy by cell fusion'. It appeared on 26 July 1969.

The first public response was a rather testy letter to *Nature* that treated the whole of our work, and especially the interpretation we had given it, with great scepticism. It was a simple matter to demolish the letter, but I knew as soon as I saw it that I was in for another long, hard battle. It was nuclear RNA all over again. But my circumstances had changed. The heterodox position I had adopted with respect to nuclear RNA had threatened my career; but I could wage war about malignancy from a position of great psychological and material security. New items of chemical information—new structures, new reactions, new linkages—are usually readily understood and quickly assimilated, perhaps because of the precision of chemical methods. New biological ideas make their way very slowly. It took a decade before my experiments on nuclear RNA were fully accepted; and more than that before it was generally acknowledged that there were indeed normal genes that could suppress malignancy.

With Francis Wiener in Stockholm doing most of the chromosome analysis, George and I hammered on, and eventually we accumulated so much evidence for the suppression of malignancy that anyone turning his attention to the subject had to take our experiments into account, whatever he might think of our interpretation of them. I was presented with an unexpected opportunity to give a public exposition of the work we had done and what I thought it meant. The Royal Society has a number of prize lectures of which the oldest is the Croonian, named after the benefactor, Dr Croone, who endowed it in 1684. Most Croonian Lecturers use the occasion to review their life's work, but since I did not feel that my life's work was done, I decided to defect from that ponderous tradition. The title of my Croonian lecture was 'Cell fusion and the analysis of malignancy', and I talked about the work on which I was currently engaged. My conclusion was that we already had ample evidence that malignancy could be suppressed by the activity of normal genes and that this was not a trivial matter. My last sentence was full of bounce: 'That malignancy can be suppressed, and suppressed by the activity of a normal body cell, is, it seems to me, no small thing; and perhaps I may be forgiven for hoping that the further exploration of this phenomenon may contribute in some small way to our understanding of what remains one of the most distressing of human maladies'.

Why was our evidence not at once accepted? Scientists are, of course, no different from anyone else in being reluctant to give up ideas to which they are attached, but, with our experiments, there was a more serious difficulty. When others tried to repeat them with different combinations of malignant and non-malignant cells, some found, as we had done, that malignancy was suppressed, but others did not. I naturally examined with great care every paper in which it was claimed that malignancy was not suppressed, and I was able to convince myself that the failure to observe the suppression was due in almost every case to the fact that the hybrid cells had already undergone substantial losses of chromosomes before they were tested. It is one thing to convince oneself; quite another to convince everyone else. To do that, I would have had to re-examine every cross for which a negative result had been reported and demonstrate decisively that my explanation for the failure to observe suppression was correct. That way lay madness. I decided instead to attempt to unravel the mechanism by which the suppression was imposed, in the hope that, if I could explain *how* suppression occurred, people would have to accept *that* it occurred.

That proved to be much more difficult than I had supposed. I began by exploring whether it might be possible to locate the genes that were doing the suppressing. This would have been a hopeless task but for a timely discovery made in Stockholm by Lore Zech and Torbjörn Caspersson. They had found that if chromosomes were stained in a particular way, they showed a pattern of bands that permitted individual chromosomes, and even parts of chromosomes, to be identified with great accuracy. This finding completely changed the character of chromosome analysis, and it made the investigation I had in mind a plausible proposition. I have already explained how the loss of chromosomes in hybrid cells can be exploited to provide a method for assigning genes to specific locations. It was a variant of this method that I now applied to the problem of locating the genes that suppressed malignancy.

For simple genetic markers that could be easily scored, this was by then a fairly routine procedure; but for a marker as complex as malignancy, the task was obviously going to be much more difficult, and the outcome uncertain. I spent three years on it with Jon Jonasson, who had come to the Dunn School from Sweden, and a further two with Ted Evans. Ted had been one of Charles Ford's collaborators in the original study that had been made on the chromosomes of the first man–mouse hybrid cells and was now a member of the Dunn School staff. In the end we were able to provide good evidence that a gene or group of genes located in a particular section of one identifiable chromosome did indeed play a crucial role in determining the suppression of malignancy.

While I was doing this, my position was gradually being strengthened by work in other laboratories. Papers reporting the suppression of malignancy in an increasing range of cell combinations continued to accumulate; and those who had originally argued strongly for the dominance of malignancy in hybrid cells became more circumspect in the analysis of their results. An important contribution was made by Eric Stanbridge, an expatriate Englishman working at the University of California at Irvine. Eric had extended our analysis to human cells and found that, in some crosses between malignant cells and non-malignant ones, loss of chromosomes occurred very slowly. This resulted in a much less transient suppression of malignancy than that seen in most of the mouse hybrid cells that we had studied. Eric later spent a sabbatical year with me at Oxford and we had a lot of fun together. Eventually it was shown that, in human hybrid cells too, a single chromosome played a decisive role in suppressing malignancy and that this chromosome carried a group of genes that corresponded exactly to those found on the chromosome that suppressed malignancy in mouse hybrid cells. The evidence for suppression had become very strong.

Ten years after I had Crooned at the Royal Society, I was invited to give the opening plenary lecture at the Second International Congress on Cell Biology in Berlin. I was asked by my hosts to combine history with science, a difficult thing to do without trivializing both. The title of my talk was 'The city of Berlin and some modern approaches to the cancer problem'. I began by reviewing the unparalleled contribution that scientists working in Berlin had made to the development of modern cell biology, and I used the cataclysmic recent history of the city to expound my views on what does and what does not constitute a propitious environment for creative science. Then I reviewed a decade of work on the analysis of malignancy by cell fusion, showed how the evidence for the suppression of malignancy by normal genes had become progressively stronger, and indicated where further exploration of the problem might go. I ended, as usual, by talking about work on which I was currently engaged.

A little later Ruth Sager, a distinguished geneticist who had moved from microbial into somatic cell genetics, wrote a review for

Advances in Cancer Research. It bore the title 'Genetic suppression of tumor formation' and ended with a summary that enumerated the principal conclusions she had reached. The first of these was: 'The original observations of tumor suppression in cell hybrids by Harris, Klein and colleagues have been confirmed and extended by many investigators using a variety of cell types'. And a bit further down the list: 'Detection of suppression in cell hybrids depends upon retention of chromosomes carrying suppressor genes'. I could hardly disagree. A couple of years after that, at the Fourteenth International Cancer Congress, there was a whole session devoted to genetic elements that suppress malignancy.

I wrote two books about cell fusion. The first was a formal version of the Dunham Lectures that I gave at Harvard. It was called *Cell Fusion* and was written before I had begun to work on malignancy. The second was called *La Fusion Cellulaire*. It was not, however, a translation of the first, but a new book written some years later in French. On the initiative of François Jacob, I was elected into a professorship at the Collège de France and gave a series of lectures there. Paris was at its beautiful best, and my hosts were the very essence of civilized hospitality. A singular reward for my services was the privilege of parking my car in the courtyard of the Collège. Those who know Paris will be aware of the violence that is done to obtain a parking-place in the Rue des Écoles; but I simply drove up to the gates of the Collège, which were opened for me, and parked my car in the shadow of Champollion's statue.

The lecture theatre echoed with historical associations. It was the room where Renan was stopped in his tracks when he uttered that memorable phrase 'Jésus était un homme admirable'. On the wall there was a bronze plaque of Bergson. Whenever I receive an invitation to give a lecture in France, I take the trouble to prepare it in French. That's a gesture of gratitude I make for the endless pleasure that the language and literature of France have given me throughout my life. As there was no book in French on the subject of cell fusion, I decided to publish the lectures I gave at the Collège. *La Fusion Cellulaire* was the result. It was, of course, much more difficult to write than *Cell Fusion* and is a better book.

Scientific writing is a very funny business. The formal constraints that editors impose and readers expect are severe. A scientific paper may be solid or spurious, elegant or barbarous, original or dull; but whatever its quality, it will make its appearance encased in a remarkably rigid and standardized framework. Writing a paper for a scientific journal is rather like doing physical exercise in a straitjacket. Scientific books permit a little more in the way of self-expression, but it is only very rarely that they offer an opportunity for creative writing. The journalistic response to our first paper on cell fusion brought me to the attention of some enterprising editors, and I began to receive invitations to write articles of a non-scientific nature. I had written nothing but scientific books and papers since my undergraduate days, and I was amazed at the delight with which I again took up my literary pen. At first there were some book reviews for the *Times Literary Supplement*; and later, a few serious pieces, on subjects in which I had a genuine interest. It didn't amount to much, but for me it was speech after long silence.

11

THE QUEEN'S PROFESSOR

In 1979 Alexandra and I took a spring holiday in Crete. The children had completed their Oxford education and had gone off to make their own ways in the world. Family summer holidays were a thing of the past. We were having lunch on the sunlit terrace of the hotel where we were staying when one of the waiters came over to say that there was a telephone call for me from Oxford. In Oxford only my secretary knew where I was, and she would not have rung me unless there was an emergency. It was therefore with some relief that I heard her say that there was no cause for concern, and that she simply wanted to know what she should do with an envelope marked *personal and confidential* that had come from 10 Downing Street. I told her to open it and read me the contents. The envelope contained two letters. One was from the Prime Minister, then James Callaghan, and the other from the Appointments Secretary in the Cabinet Office. The Prime Minister wanted to know whether I would agree to his recommending my name to the Queen for appointment to the Regius Professorship of Medicine at Oxford. The Appointments Secretary, aware that the Prime Minister's proposal would create difficulties for me, hoped that I would none the less explore the situation with the Vice-Chancellor at Oxford, with whom he had apparently already discussed the matter.

The Regius Professorship of Medicine at Oxford is a piece of English history. It was founded in 1546 by Henry VIII and is an appointment made by the monarch, in recent times on the recommendation of the prime minister. The Appointments Secretary of the Cabinet Office takes soundings at Oxford and elsewhere in the country and then presents an annotated short list of names for the prime minister to choose from. The process of consultation is conducted with great confidentiality and, although rumours abound, it is usually very difficult to discover just how or why a particular choice is made. In my case, the choice was very surprising. I was known to be a laboratory scientist still at the bench, and I had had little experience of the world of clinical medicine. The great majority of those who had held the position in the past had been distinguished clinicians, and I do not think any of them was an active experimentalist at the time he was appointed. I could imagine, in a moment of conceit, that my scientific colleagues in South Parks Road might have suggested my name to the Appointments Secretary; but he must also have consulted clinical opinion in Oxford, and it was difficult to believe that my name could have gone forward without some support from the clinicians.

I had never taken much trouble to disguise my views on clinical research. I believed that it had to be judged by the same exacting criteria as laboratory research and that, by those criteria, much of what went on in the clinic was trivial, if not worse. These were not views that were likely to endear me to my clinical colleagues, and I still regard it as remarkable that some of them should none the less have thought that I did have a contribution to make to clinical medicine in Oxford. However, I thought it very unlikely, when I received the Prime Minister's letter, that I would be able to accept the position, signal honour though it might be. The main, and at first sight insuperable, difficulty was how, as Regius Professor, I could continue with my experimental work. This was not so much a question of whether my duties would leave me time for research (I had long ago learned how to manage my time), but where this research could be located.

If I gave up the Professorship of Pathology, the convention was that I would have to leave the Dunn School in order to avoid being an embarrassment to my successor. This would mean giving up all the services and facilities that I had built up over the years and that were essential for my work. To duplicate these elsewhere, even if space could be found, would be very expensive, time-consuming, and wasteful. I therefore had little hope, when I went to see the Vice-Chancellor, that any arrangement could be made that would enable me to accept the Regius Professorship without curtailing my experimental work, which I was not under any circumstances prepared to do. The Vice-Chancellor at that time was Rex Richards, whom I had first met when I was a D.Phil. student at Lincoln. I did not have to explain the difficulty to him. We mulled over the possibilities and finally settled on the only one that seemed to me to stand a chance.

I have already explained that at Oxford you can do what you like if you can carry your colleagues with you. The scheme that Rex and

I decided to try out was, I suppose, a constitutional outrage, but it was just possible that the executive bodies of the University might wear it if they were convinced that some academic advantages might result and that no injustice would be done. I wrote a holding letter to the Prime Minister and then prepared a paper for consideration by the two relevant faculty boards, Physiological Sciences and Clinical Medicine. What I proposed was that the Chair of Pathology should be suspended and that the Dunn School should be allocated to the Regius Professor of Medicine for so long as I held that position. This scheme would permit me to remain the head of the Dunn School, and I would then be able to pursue my research as before. I undertook to continue to do whatever was expected of me as head of the Department.

But what of the duties of the Regius Professor of Medicine? Well, it was rather difficult to ascertain just what these were. The statutes were couched in quite general terms and imposed no specific burden. Previous Regius Professors, at least during the last century or so, had done many different things. Osler had been a practising clinician, Burdon-Sanderson a physiologist; Garrod had founded the science of human genetics. I set out for the faculty boards what I thought I could do and what I couldn't do, and then left it to them to decide whether the package was acceptable or not.

It took the best part of a term for the two boards to come to a decision, but they both eventually agreed to accept my suggestion. The scheme then went to the General Board of the Faculties and the Hebdomadal Council. These bodies were exercised about other aspects of my proposal, such as whether it constituted a form of professorial pluralism to which they were in principle opposed. Since there was no question of pluralism of stipend, only pluralism of duties, this did not seem to me to be an unsurmountable objection. I must have had some staunch advocates on the Hebdomadal Council for it is a very conservative body. In the end, it did, however, agree to promote a decree to suspend the Professorship of Pathology and allocate the Dunn School to the Regius Professor of Medicine. I wrote to the Prime Minister and agreed that my name could go forward.

Some three months later I received a Letter Patent from the Queen. It came in an elegant red morocco leather envelope and was couched in beautiful Tudor language. Of her 'especial grace certain knowledge and mere motion', Queen Elizabeth the Second had granted unto her 'trusty and well beloved Henry Harris . . . the Reading and Office and Place of Our Reader in the Science and Faculty of Physic for the use and improvement of the English youth in Our University of Oxford'. Thus I became the twenty-ninth Regius Professor of Medicine.

It might well be asked what I hoped to achieve in this new role. The Regius Professor has no formal executive authority, but if he has the confidence of his colleagues, he can exercise a great deal of influence in Oxford. He is a member of all professorial electoral boards both in the medical sciences and in clinical medicine; and he sits *ex officio* on the key committees that determine academic policy in these fields. The medical science departments in South Parks Road did not pose great problems; but in the rapidly growing clinical school there was an important issue to be settled. My youthful experiences had left an indelible memory of what a clinical school ought not to be; and I knew what qualities a clinical school would have to have at Oxford if it was to achieve a respectable place in the life of the University. In a clinical environment there is always a tug of war between those who regard good doctoring as the highest goal and those who seek to inject some science into the hospital scene. There is competition for resources, for positions, and for the minds of the young. In the United Kingdom, it is the good doctors who have on the whole won out. But in Oxford, being a good doctor was simply not enough. If you wanted to be regarded as an academic amongst academics, then your life had to have some intellectual content in addition to that provided by the exercise of professional clinical skills.

Within the framework of the National Health Service, which imposes huge service responsibilities on its doctors, it is very difficult for a clinician to find time for serious experimental work, even if, at some stage in his career, he has had the opportunity to acquire the necessary training. If the aim is to inject some professional science into the hospital scene, then one must inject some professional scientists. However, good professional scientists will not work (at least not for long) in a department where the head does not know one end of an experiment from another. So the first requirement is that the clinical professor should be someone who has actually made a serious contribution to science; and the second is that he should be provided with some posts for professional scientists within his department.

Neither requirement is easy to fulfil, especially when times are hard. Still, we haven't done too badly at Oxford. I made it clear at the outset that, in so far as it lay within my power, no one would be appointed to a professorship at Oxford simply because he was a good doctor; and it has turned out in fact that, in recent years, all the appointments to clinical chairs have gone to men who have had a substantial first-hand experience of experimental science. What's more, we have managed, in one way or another, to provide them with positions for professional scientists. Not as many as we would like, but at least something to begin with. If one can judge by the response of the Medical Research Council and the other agencies that provide funds for the advance of medicine, or by the quality of the students who apply for clinical places at Oxford, the experiment is succeeding. And incidentally, the doctoring gets better and better.

There is one duty that is laid on the Regius Professor not by statute, but by royal decree. In 1617, on the prompting of James I, whose motive was probably to avoid paying the Regius Professor an appropriate stipend, the holder of the post was established as Master of the Hospital of Ewelme, 'for the professor's better sustenation even though he be a mere layman and have not taken orders'. The Hospital (hospice) at Ewelme is an almshouse which was founded in 1437 by William De La Pole, Earl of Suffolk, and his wife, Alice, the granddaughter of Geoffrey Chaucer. The original endowment was for the support of two chaplains and thirteen poor men, and the first master was appointed in 1442. I am the thirty-eighth master.

Alice, who lies buried in the church at Ewelme, left large holdings of agricultural land for the support of the almshouse, and the revenue from these estates is still used for this purpose. The almshouse is built as a cloister around a small courtyard that lies between the Church and an adjoining schoolhouse which was also endowed by Alice (Plate 17). The Church, the almshouse, and the school still function as originally intended, and together they present a remarkably evocative picture of English life five hundred years ago. The Ewelme Trustees, who administer the almshouses and the estates that support them, make the duties of the master unexacting; and the 'sustenation' proposed by James I now takes the form of a graceful apartment that is placed at the Master's disposal. It overlooks the old courtyard, contains no telephone, and is an ideal retreat from a noisy and busy world. No rational forecast of my career could have envisaged that I would one day be responsible for the welfare of a small community of people who towards the end of their days find

Plate 17. The almshouse at Ewelme.

themselves in need of what the Hospital of Ewelme has to offer. This duty of the Regius Professor brings a quite different kind of reward.

All this naturally landed me with a bigger load of committee work at Oxford than I had had before, but I offset this by shedding even more of my London commitments and by travelling even less. Although some of my colleagues laid odds against it, I remained immersed in my experiments. For reasons that I do not myself fully understand, I began to prefer to work alone. The members of the Cell Biology Unit, the D.Phil. students, the overseas visitors were still there, and of course I continued to collaborate with them. The studies with hybrid cells had revealed a number of interesting and hitherto unsuspected features of tumours, and these we explored together. But there was one problem that I reserved for myself, a war of attrition that I waged against the cancer cell in private. Although the suppression of malignancy by normal genes was now an established fact, and although we had some understanding of the chromosomal basis of the phenomenon, we still didn't have the foggiest idea of how, in biochemical terms, the suppression was brought about. I was convinced that we wouldn't get to the bottom of the cancer problem until we had unravelled the fail-safe device that held malignancy in check. In a slightly obsessional fashion,

I picked away at this from many different angles, but I made little headway.

Eventually, as I neared my sixtieth year, I turned, perhaps in desperation, to a method of investigation that I disliked: conventional histopathology. Histopathology is the microscopic study of diseased tissues. Samples of such tissues are taken from the body (usually after death), treated with a fixative, and cut into thin sections. These can then be stained in various ways and examined under the microscope. This is a routine procedure for diagnostic purposes, but, for the experimentalist, it has the profound drawback that it presents the observer with no more than a static picture of events, and causal relationships can only very rarely be deduced. None the less, there was essentially no other way of looking at the lesions produced in the tissues by the hybrid cells that I injected; and no doubt I would have looked years earlier had I not been so reluctant to rely on histopathological techniques. The hybrids that I now set out to examine in this way were crosses between different kinds of cancer cell and normal fibroblasts.

I have already explained what fibroblasts are. They are the cells that are primarily responsible for the repair of a wound. They move into the wounded area, multiply there, and secrete a number of substances that eventually give rise to scar-tissue. These substances form a fibrous web, called in the jargon an extracellular matrix, and as this web builds up and condenses around the cells, they stop multiplying. When I examined the lesions produced by the injection of hybrids in which malignancy was suppressed, I found that they behaved like normal fibroblasts: they produced the fibrous extracellular matrix and stopped multiplying. Within the hybrid cell, the fibroblast had imposed its normal pattern of differentiation on the cancer cell. But when I looked at hybrids which, because they had eliminated the relevant chromosomes, had become malignant again, I found that they did not produce a fibrous extracellular matrix and did not stop multiplying. I explored this finding very thoroughly, and I came to the conclusion that if the hybrid cell behaved like its normal fibroblast parent and produced the characteristic fibrous extracellular matrix, then it would not multiply progressively in the host into which it was injected. The suppression of malignancy was linked to the production of the extracellular matrix.

My thoughts then moved from the particular to the general, and in the process my ideas about cancer were turned upside down. The two great unscaled peaks of cell biology, cancer and differentiation, merged into one. Differentiation, as I have already mentioned, is the general term used to describe the process by which the embryo develops different organs and different families of specialized cells. To begin with, the egg fertilized by the sperm multiplies rapidly and generates a homogeneous ball of cells, indistinguishable one from the other. However, as the embryo develops further, areas of specialization arise, and within these areas the different tissues of the body are formed. Differentiation gradually brings the unrestrained multiplication of the embryonic cells under control, and, in some tissues, cell multiplication is brought to a complete stop. For the fibroblast, differentiation is completed when cell multiplication has stopped and the fibrous extracellular matrix that the cells have produced has condensed into a scar. What I was seeing in the hybrids in which malignancy was suppressed was the process of differentiation going to completion. And where the hybrid cell had lost the ability to produce the fibrous extracellular matrix, and was thus unable to complete its differentiation, cell multiplication continued and a progressive tumour was formed.

For my sixtieth birthday, I wrote myself a paper entitled 'Suppression of malignancy in hybrid cells: the mechanism' and sent it off to the *Journal of Cell Science* in which most of my work on the analysis of malignancy had been published. The *Journal of Cell Science* is a conservative journal, and for most of the paper I stuck diligently to what I had actually observed. But in the discussion, I let myself go. I argued that cells were so constituted that it was perfectly natural for them to multiply continuously and to do so until their multiplication was brought to a stop by differentiation. If, for any reason, the appropriate form of differentiation could not take place, as might happen, for example, if a cell sustained mutations that impeded the process, then one should expect cell multiplication to continue unrestrained. On this view, there was nothing fundamentally abnormal about the multiplication of cancer cells. What was abnormal was that they had lost the ability to bring their differentiation to its normal conclusion. I have complete confidence that these ideas will be energetically resisted. The task of reducing them to biochemical terms will be formidable, and it will no doubt see me out. But so long as hand and eye and brain remain intact, I intend to plug away at it (Plate 18).

Christ Church is the college to which the Regius Professorship is attached, and when I was appointed I had to leave Lincoln to become

Plate 18. The clear light of day in the Professor's laboratory.

a Student of The House. Christ Church is almost always referred to in Oxford as The House, and its fellows are, for curious historical reasons, known as Students. Lincoln is not a big college, and it manages to achieve an intimate, one could even say domestic, atmosphere. Over the years, I had come to know almost everyone there very well. Christ Church, on the other hand, is the largest and grandest of all Oxford colleges. It has magnificent buildings, the most celebrated meadow in England, a famous collection of paintings. It is the college of Thomas More, Philip Sidney, Robert Burton, John Locke, 'Lewis Carroll', and a forbidding array of prime ministers and viceroys of India whose portraits grace (or disgrace) its walls. For a brief period it was also the college of Albert Einstein. When I left Lincoln, the College kindly elected me to an honorary fellowship, so I am now able to partake of grandeur or domesticity as the mood takes me. When they reach a certain age, professors are sometimes asked whether they would allow their names to go forward for possible election to the headship of a college. I received three approaches of this kind, but my heart was in the laboratory, not in college life, and I let these flattering opportunities go by.

At the beginning of last term, I went into the undergraduate class to give my historical lecture for the twenty-second time. Out of curiosity, I asked all those who had heard of Howard Florey to put up their hands. In a class of one hundred second-year medical students and at Oxford, six hesitant hands were raised. In a few years from now, if the fates are kind, I shall retire. My successor will sit at the desk where I am now writing these lines and will do his experiments in the adjoining laboratory where Florey tested penicillin and I fused cells. If one day, out of curiosity, he were to go into the undergraduate class and ask all those who knew who Henry Harris was to raise their hands, would a single hand be raised?

> We may return to Mozart.
> He was young, and we, we are old.
> The snow is falling
> And the streets are full of cries.

INDEX

Abercrombie, M. 166
Abraham, E. P. 63, 80, 82, 173, 178
Acetabularia 104, 124, 132, 144, 146, 148
  mediterranea 145
  phosphates in 149
achlorhydria 98
actinomycin D 153, 177, 178
Adelaide, Australia 53
Adelberg, E. 203
'Advance Australia Fair' 3
Advances in Cancer Research 222
Agricultural Research Council 113, 114, 160, 168, 213
  Secretaryship 214
Albury, N.S.W. 48
Alden, Nellie 60, 74
Alden, Priscilla 60
Alexander, Samuel 27
Alice, Duchess of Suffolk 228
amino acids 96
anatomy 34, 35
Anderson, John 24, 27, 72
Andersonians 28
animal species 187
Annapolis, USA 130
anucleate cells 124
Anzacs 17
Appointments Secretary, Cabinet Office 224
Arna 33
arteriosclerosis 99
Arts undergraduates 24
atheism 11, 27, 128, 164
atomic bombs 46
Atomic Energy Research Establishment, Harwell 97
Auburn Public School 1
Australia 77
Australian aborigines 18
Australian bush 17
Australian Legends 18
Australian National University 45, 51, 63, 89, 100, 210
  Vice-Chancellorship 209
auto-immune disease 40

autoradiographic techniques 189
autoradiographs 188
autoradiography 106
Avery, O. T. 105

β-galactosidase 146
Bacillus cereus 178
bacteria 158
bacterial enzyme 146
bacteriology 37
Baltimore and Ohio railway 134
Balwyn, Melbourne 49
Balwyn Road, Melbourne 49
Barclay, W. R. 95, 134, 205
Barski, G. 191, 197, 200
Basil Blackwell's bookshop 71
Bateson, W. 113
baths 165
Bayfordbury, Hertfordshire 112, 138, 161
Beadle, G. W. 111, 135, 136, 157
  Muriel 111
Beg-Meil 184
Belgium 60
Benditt, E. P. 202, 204
Bennet-Clark, T. A. 160
Benoit, J. 105
Berenblum, I. 82
Bergson, Henri 222
Berkeley, California 135
Berlin 211, 221
Berlin Alexanderplatz 33
Beth Israel Hospital, Boston 170
Bethesda, Maryland 119
Biochemical Journal 106, 107, 149
Biochemical Society 147
biochemistry 34, 35
Biographical Memoirs of Fellows of the Royal Society 105
Blackburn, C. B. 43
Blackburn, R. B. 43
blood chemistry 43
Blue Mountains, N.S.W. 18
Bollman, J. L. 99
Bondi Public School 2

INDEX

botany 34, 132
Botany Bay, N.S.W. 18
Bragg, W. L. 91
Bramwell, M. E. 148, 170
Brayden, H. 20, 25
bread mould 157
Bremer, H.-J. 149
Brennan, C. J. 26
Brenner, S. 71, 148, 154
British Empire Cancer Campaign 90, 96, 116, 169
British Journal of Experimental Pathology 94
British Navy 17, 22, 25
Brittany 184
Brno 198
broad bean 151
Broad Green Wood, Bayfordbury 114, 137, 138
Brodsky, Alexandra 46
Brokaw, C. J. 79
Brown, Dr John 33
Brozzu, G. 80
Brussels 61
Bryson, V. 182
Burdon-Sanderson, John 226
Burnet, F. M. 42, 51
Burragorang Valley, N.S.W. 19
Bush, S. 70
bush ballads 4

Cabinet Office 79
California Institute of Technology 79, 136
Callaghan, James, Prime Minister 224, 226
Canberra 45, 77, 100, 209, 210
cancer 80, 82, 215, 231, 207 *et seq.*
  recessive mutations in 217
cancer cells 87, 92, 230
  abnormality in 87
  breast 187
  multiplication of 231
  uterine 187
Cancer Institute, Melbourne 101
Cancer Research Campaign 169, 208
  Cell Biology Unit 169, 170, 173, 229
  Scientific Committee 213, 214
carbon dioxide 92
Carnegie Institution of Washington 97
Casperson, T. 220

cell biology 100, 203
Cell Fusion 222, 185 *et seq.*
cell fusion 159
  virus-induced 199
cell multiplication 80, 103, 231
cell nucleus 107, 140, 155
cephalosporins 80
Cephalosporium 80
Chamberlain, Neville 17, 21
Champollion, J.-F. 222
Chansons de Geste 26
Chaucer, Geoffrey 228
Chekhov, Anton 33
chemistry 34
chemotaxis 61, 66, 68, 77, 78, 85
Chesapeake Bay 130
Chevrolet 122
Chibnall, A. C. 91
Chicago 135
  Stockyards Inn 135
Children's Cancer Research Foundation, Boston 198
chimaeras 193
Chinatown, San Francisco 135
Chionodoxa 55, 77, 163
chloroplasts 151
Christ Church, Oxford 231
  students of 232
chromosome banding 220
chromosomes 182, 190, 202
  human 190
  mouse 190
Chu, E. H. Y. 203
Churchill, Winston S. 21
cilia 79
cinematographic records 197
Clarendon Press, Oxford 206
clinical research 225
Cohen, P. P. 110, 204
Cohn, H. 136
collaboration 218
Collège de France, professorship 222
colon bacillus 97, 142
Columbia University 217
Comandon, J. 66
committee work 214
complementation 157
Condroz 60
conformism 128
Conlon, A. 42
Cook, E. W. 112
Cook, P. R. 200

INDEX 237

Coombs, H. C. 209
Copland, D. B. 66
Coral Sea battle 31
Cornefert, Françoise 191
Cotswolds 74
Coventry Evening Telegraph 193
Cox, E. G. 168
Craigie, J. 109
Crete 223
Crick, F. H. C. 76, 91, 148, 154
Croonian Lecture 219
Curtin, John, Prime Minister 25
customer-contractor principle 79
Cutler, A. R. 22
Cynamid Company 112
Cynamid European Research Institute 112
cytology 100
cytoplasm 104
Czechoslovakia 17

Darlington, C. D. 113, 173
Darnell, J. E. 125, 181
Darwin, Charles 34, 84, 85, 182
Daudet, Alphonse 17
Daughters of the American Revolution 118
Déak, I. 200
Dean, H. R. 151
Decameron 28, 30
Deigma 29
Delaware 119
Delbrück, M. 136
Denton, D. A. 50
Denver, Colorado 137
Descent of Man 19
Dewey, John 118
dietary fats 99
differentiation 230, 231
DNA (deoxyribonucleic acid) 76, 91, 105, 149, 155, 182, 195
  not coding for protein 182, 183
  precursor 188
  replication of 195
Döblin, Alfred 33
Dodds, E. C. 90
Dodds, K. S. 113, 137, 156, 160, 161
Dodson, L. F. 63
dogma 187
Double Bay, Sydney Harbour 32
Dr Moreau 194

Dreyer, G. 167
Drosophila 201
Dubos, R. 133
ducks 40, 105
Dulbecco, R. 136
Dunham Lectures, Harvard 203, 207, 222
Dunn School Mafia in the USA 203, 204, 205
Dynamic State of Body Constituents 102

Eagle, H. 125
East St Kilda, Melbourne 51
Ebel, J.-P. 156
Ebert, R. H. 204
eccentricity in Oxford 129
Eccles, J. C. 37
Edström, J.-E. 181
egg 231
Ehrlich, Paul 69, 84, 85
Ehrlich ascites tumour 187
Ehrlich cells 188
Einstein, Albert 84
Elkind, M. M. 120, 124, 136
Ellmann, R. 132
Engel, E. 203
Engelsberg, E. 203
English countryside 74
English land law 172
English teaching 21
Ennor, A. H. 100
enucleated cells 149
Ephrussi, Anne 198
Ephrussi, B. 191, 195, 197
Escherichia coli 97, 102, 142, 146, 202
European Molecular Biology Laboratory, Scientific Adviser 213
Evans, E. P. 220
Evans-Pritchard, E. E. 71
evolutionary change 182
Evolving genes and proteins 182
Ewelme almshouse 228, 229
Examinations
  Intermediate Certificate 13, 17, 19
  Leaving Certificate 19
  Qualifying Certificate (QC) 8
Experimental Cell Research 159
extracellular matrix 230, 231

INDEX

Faraday, Michael 84
fermentation 92, 94
fibroblasts 87, 88, 89, 93, 94, 230
Fildes, P. G. 63, 64, 69
Fincham, J. R. S. 157
Fisher, H. W. 140
flagella 79
flame photometer 43
Fleming, Alexander 78
Florey, Howard 44, 52, 54, 57, 58, 61, 62, 67, 74, 80, 86, 91, 98, 151, 152, 162, 164, 166, 192, 206, 233
Florey, Mary Ethel 64, 77, 206
Fontamara 29
Ford, C. E. 190, 195, 196
France 184
Fredericksburg 130
Fremantle, Australia 53
French, J. E. 98, 99
French studies 23, 27
Freud, Sigmund 27
frog's bladder 94
Fulton, J. F. 36
fungi 157
fused cells 188
  division of 190

Gamble, J. L. 44
Garrod, Archibald 226
gas exchanges 93
gene 182
gene regulation 181
gene switches 141, 142, 146, 182
gene-mapping 201, 202
General Pathology 82
genes 141
  dominant 215
  of man 188
  of mouse 188
  recessive 215
  regulatory 141
  replicated 189
  that suppress malignancy 219, 221
  transcribed 189
genetic code 183
genetic defects in malignancy 218
genetic exchanges 159, 187
genetic information 145
genetic instructions 155
genetic markers 186, 196
genetic operator model 142, 149

genetic specifications 141
genetical analysis 191
genetical methods 158, 199, 201
genetics 132
German studies 23, 27
Germany 210
Gey, G. O. 97
'giant' cells 159
Gilbert, N. E. 153
Gilbert, W. 184
Gladstone, G. P. 63
glutamic acid 157
glutamic dehydrogenase 157
Golden Lotus 30
Goss, S. J. 202
Gowans, J. L. 63, 78, 98, 99, 114, 169, 173, 194
Gray, J. 208
Great Australian Bight 53
Green, H. 201
Greenstein, J. P. 87
gum trees 169, 171
gynaecology 39, 41

H. E. Waldron Memorial Prize 39
haemagglutinating virus of Japan 159
haemoglobin 146
Haldane, J. B. S. 76, 113
Hämmerling, J. 104, 124, 144, 146, 147, 198, 211
Hancock, W. K. 77
Harley, J. L. 162
Harris, Alexandra 48
Harris, Ann 96, 138
Harris, Helen 96, 138
Harris, Paul 52, 74, 138
Harris, R. J. C. 110
Harris Brothers 1, 6
Harvard 208
  chair 207
Harvey, William 84
Hawkesbury River, N.S.W. 19
Hawthorn, Melbourne 52
Heatley, N. G. 63
Heiss, J. 212
HeLa cells 97, 187, 188
Henry VIII 224
Heppel, L. A. 125
herpes virus 185
Hertford 138
heterokaryons 158, 159, 160, 188, 189, 195, 197, 200

INDEX 239

Hiatt, H. H. 170
Himsworth, H. P. 167, 208
Hinshelwood, C. N. 71
histopathology 230
historicism 30
history of science 172
Hitler's assumption of power 15
Hospital of Ewelme, Master of 228
housing shortages 49
human chromosomes, elimination of 202
Hume–Barbour Trophy 13
Huy 60
HVJ virus (Haemagglutinating Virus of Japan) 185
hybrid cells 189, 191, 201
  colonies of 197
  mononucleate 197
hybrid cells derived from mouse and man 191
hybrids, mouse-cell 192
hybrids between malignant and normal cells 218
hyphae 157
  fusion of 158, 186

immunological reactions 186
immunology 37
Imperial Cancer Research Fund 109, 110
Innes, John 113
Institut Gustav Roussy 191
Institute for Cell Biology, Dahlem 211
Institute for Medical Cell Research and Genetics, Stockholm 200
Institute of Microbiology, Rutgers University, New Jersey 182
International Cancer Congress (XIVth) 222
International Congress for Cell Biology (IXth) 103
International Congress of Biochemistry (Vth) 147
International Congress on Cell Biology (IInd) 221
Invisible man 4
Italian studies 23, 27
itinerant preachers 198

Jacob, F. 102, 141, 144, 202, 222
Jahnz, Marianne 97, 123, 137, 138, 147, 148, 170

James I 228
Jamestown, Virginia 130
Japan 22
Japanese air force 25
Jennings, Margaret A. 63, 206
Jervis Bay, N.S.W. 48, 171
John, Augustus 91
John Curtin School of Medical Research 45, 77, 95, 98, 100, 102
  Directorship of 209, 210
John Innes Horticultural Institution 112
John Innes Institute 156, 209
  Department of Cell Biology 113, 114, 115, 152, 161
Johnson, R. T. 201
Jonasson, J. 220
Jones, E. R. H. 173
Jones, F. S. 88
Journal of Cell Science 231
Journal of Molecular Biology 141, 142
Joyce, James 27, 132

Kanematsu Institute, Sydney 37
Karolinska Institute, Stockholm 181
  Department of Tumour Biology 218
Katoomba, N.S.W. 18
Katz, B. 37
Kay, D. 64
Kelley, W. N. 203
Kendall, Henry 4
Kennedy–Nixon debates 133
Kensington Public School 2
Kent, J. H. D. 65, 206
kidney failure 43
Killip, Jas. A. 14
Klein, G. 218
Klinger, H. P. 203
Koch, Robert 84, 85
Krebs, H. A. 69, 76, 84, 91, 92, 110, 112, 166
Kuffler, S. W. 37

La Dernière Classe 17
La Fusion Cellulaire 222
Lacks, Henrietta 97
Lacour, L. F. 150
lactic acid 92, 95

INDEX

lactose (sugar of milk) 146
Lambie, C. G. 39
Lancieux 184
Landi, Elissa 4
Latin 20, 21
Lavoisier, Antoine 84
Lawson, Henry 4
Le Gros Clark, W. E. 71
Le Petit, M. 28
Le Soir 193
Leaf, A. 94, 133
Leckie, B. D. 179
Les Landes 184
Letter Patent 226
leucocyte traces 67
leucocytes 61, 66
leukotaxine 62, 78, 80
Lewis, E. B. 136
liberal education 25
Life magazine 67
Lincoln College, Oxford 65, 175, 232
 middle common room 66
 Rector's lodgings 66
Lister Institute, Governing Body 213
Littlefield, J. W. 134, 197, 203, 205
liver cells 170
Loftfield, R. B. 134
logical positivism 29
London *Daily Mirror* 192
lymphocytes 78, 99, 211
Lynen, F. 211

McArdle Memorial Laboratory for Cancer Research 111
McCarty, M. 105
McClintock, Barbara 181
McCormick, Peter Dodds 3
McDonald, I. R. 50
Macfarlane, R. G. 75, 80, 85
McIntyre, A. K. 37
Mackaness, G. B. 63, 97, 132, 205
Mackellar, Dorothea 4
McLeod, C. M. 105
macrophages 68, 87, 94, 107, 108
malignancy
 mechanism of suppression 231
 suppression of 218, 219, 221, 229, 230
malignant cells 215
Man on his Nature 76
man–mouse hybrid cells 195, 201
Manhattan, New York 117

Marchesi, V. T. 205
Marrian, G. F. 109
Marston, R. Q. 204
Martin, D. W. 203
Marx, Karl 27
Maryland, USA 127
Massachusetts Avenue, Washington 127
Massachusetts General Hospital 94, 134
 Ether Dome 134
Max Planck Society 211
Medawar, P. B. 91, 112, 115, 186, 195, 209
Medical Journal of Australia 41, 42
Medical Research Council, London 167, 228
 Clinical Research Centre 167
medicine 39, 41
Mediterranean coast 184
medium, selective 197
Melbourne, Victoria 49
Mellanby, E. 91, 167
Mendel, Gregor 34, 84, 85, 113, 186
Menkin, V. 62
Meselson, M. 136
Mider, G. B. 111, 112, 120, 131
Midway battle 31
military service, exemption from 24
Mill Hill 160
Miller, O. J. 203, 217
Ministry of Agriculture, Fisheries and Food, Chief Scientist 214
Mirsky, A. E. 181
mitochondria 151
modern languages 25
Monod, J. 102, 106, 141, 143, 146
Montreal 181
Mozart 21, 171, 233
multinucleate cells 158, 188
Munich Agreement 17
muscle 158
mutations
 dominant 215, 216
 recessive 215, 216

Nabokov, Vladimir 117
National Cancer Institute, Bethesda, Maryland 111, 123
National Health Service 126, 136, 227
National Institute for Medical Research, Mill Hill, London 208, 209

INDEX

National Institutes of Health,
  Bethesda, Maryland 119, 120, 121, 126, 209
  Clinical Center 120
natural selection 33
Nature 103, 142, 153, 184, 192, 195, 218
Naval Lodge, Jervis Bay 48
Nell, Marilyn B. 198
Neurospora crassa 157
New England, autumn in 130
New Jersey 119
New York 117
  Governor Clinton Hotel 118
New York Times 195
New York University School of Medicine 201
Newton, G. G. F. 63
Newton, Isaac 84, 85
Nibelungenlied 26
Nicholson, G. G. 26
Nirenberg, M. W. 125, 147
North Oxford Landlady 59
Norwich 160, 161
nuclei
  chick red cell 200
  fusion of 190
  human 190
nucleus 104
  transfer of 192
Nucleus and Cytoplasm 206

Oberling, C. 50
obstetrics 39, 41
Ogston, A. G. 107
Okada, Y. 159, 185, 188
Oléron 184
Origin of Species 19, 33
Osler, William 226
ovum 158
Owen, R. D. 136
Oxford 45, 54 *et seq.*, 163 *et seq.*
  autumn in 130
  Beech Road, Headington 59
  clinical medicine in 225
  Cumnor Hill 170
  Eagle and Child 71
  Elliston & Cavell 96
  Headington 59, 73
  Lamb and Flag 71
  Magdalen College 73
  Old Marston 76
  Radcliffe Square 73
  Rotha Field Road 96
  St Giles 61, 71
  The High 73
  Victorian houses 59
Oxford Atlas of the World 52
Oxford colleges 69
Oxford Railway Station 54
Oxford Times 54
oxygen 92, 94

P and O boats 52
Pasadena, California 136
Pasteur, Louis 84, 85
Pathological Society of Great Britain and Ireland 78
pathology 37
patronage 90
Patterson, A. B. (Banjo) 4
Pattison, Mark 167
Pearl Harbour 22
Peck, C. W. 18
'peer-review' system 213
penicillin 44, 62
penicillinase 178
Pennicuick, R. 5
Pennsylvania Station 119
Pennsylvania, USA 119
Pereira, H. G. 185
Perth, Australia 53
Perutz, M. F. 91
Peters, R. A. 70
pharmacology 37
phosphates 149
physics 34
Physiological Reviews 80
physiology 34, 35
Physiology of the Nervous System 36
Planck, Max 84
Pontecorvo, G. 159, 181, 192, 196
Poole, J. C. F. 99
popular press 193
'Poseurs' Push' 29, 30, 33
Postgraduate Medical School of London 162
Potter, V. R. 111
Poynton, Winifred 56, 83, 165
Prague 198
Prince of Wales 25
Progress in Nucleic Acid Research 147
protein synthesis 177

protein turnover 102
proteins 96, 102
Provence 184
psychoanalytic theory 30
psychology studies 23, 24
Puck, T. T. 137
Punt Road, Melbourne 51

Qantas 179
Queen Elizabeth 115
Queen Elizabeth the Second, Letter Patent from 226
Queen Mary 137

Rab and his Friends 33
Rabinowitz, M. 120, 124
radiation 125
radioactive compounds 97
Raj 53
rat–mouse hybrid cells 196
realism 27
red blood cells 43, 124, 194
red cell nucleus, reactivation of 195, 199
red cells, nucleated 195
refugees 15
registered medical practitioner 46
Regius Professor of Medicine 224 *et seq.*
  duties of 226
  powers of 227
Renan, Ernest 222
rent rackets 49
Repulse 25
respiration 92
reticulocytes 124, 146
Rhadakrishnan, S. 71
Rhodes Scholarships 72
ribosomes 141
Richards, A. R. 98
Richards, R. E. 225
Riddle of Cancer 50
Riley, W. T. 148, 170
Ringertz, N. R. 200
RNA (ribonucleic acid) 106, 108, 149, 182, 195
  breakdown of 107, 131, 143, 177
  cytoplasmic 142, 149
  flow of 200
  fractionation of 139
  in chromosomes 151
  in cytoplasm 141
  in ribosomes 142
  intranuclear breakdown of 108, 184
  intranuclear turnover 153
  messenger 142, 143, 144, 154, 155, 156, 183
  nuclear 140, 142, 149, 153, 182, 183, 199
  precursors 155
  ribosomal 154
  short-lived 131, 139, 140, 144, 147, 148, 177, 183
  synthesis 143, 178
  transcripts 184
  'transfer' 133
  turnover 106, 131, 160, 178, 181, 184
Robbins Report 160
Robinson, D. S. 99
Robinson, R. 70
Rodgers, A. 148
Rothschild, N. M. V. (Lord Rothschild) 78, 213
Rothschild Report 213
Rous, P. 88, 133
Royal Prince Alfred Hospital, Sydney 29, 36, 39, 41, 45, 179
  Clinical Research Ward 43
Royal Society 219
  Biographical Memoirs of 207
  Council 213
Rubin, H. 135
Rutherford, Ernest 87
Ryle, G. 71

Sabath, L. D. 178
Sager, Ruth 221
'Salvarsan' 69
San Francisco, California 135
Sanders, A. G. 63
Sanger, F. 91
Schneeberger, Eveline E. 200
Schnitzler, Arthur 33
Schoenheimer, R. 102
Schweiger, H.-G. 145, 149, 170
science, support of 207
Science Research Council 213
scientific writing 222
Scott, J. P. 134, 141
sea-urchin eggs 93
Seattle, Walker–Ames Visiting Professor 202

INDEX 243

Seegmiller, J. E. 203
segregated apartments 128
Sendai virus 185, 191, 194, 196
Shack, J. 120, 124
sheep 50
Shelley, P. B. 74
Sherrington, C. S. 76
Sidebottom, E. 200
Silone, Ignazio 29
Silver Spring, Maryland 122
  Rosemary Hills apartments 121
  Rosemary Hills School 127
Simpson, A. W. B. 173
Singapore, collapse of 25
Sinsheimer, R. L. 136
Sir William Dunn School of
    Pathology 38, 55, 56, 68, 168
  allocation of site 173
  extension 174
  High Court Order 173
  Professor's laboratory 164, 232
  Professor's study 55, 57, 163
  Professor's bathroom suite 164
Slater, W. K. 114, 168
slide-holder 70
Smith, Wilson 152
smog 136
sodium transport 94
somatic cells 158, 191, 199
Sorieul, S. 191
Sparrow, John 21
Spencer, T. 148, 170
sperms 79, 158, 231
  bracken 79
Spirit of Progress 48
St Andrews, Scotland 104
Stadler, V. 26, 27, 31
Stanbridge, E. J. 221
Stanford, California 135
Statue of Liberty 117
Stock, Dora 171
Stockholm 218
Stratheden 52
Stroud, H. J. 165
Struth Exhibition 34
suitcases 13
  Cordite 15, 16
sulphur 97
surgery 39, 41
Swan river, Perth 53
Sydney
  Advance Bookshop 30
  Anzac Parade 9

Avoca Street 2
Bondi 1
Bondi Beach 2, 5, 180
Bondi Road 2
Botanical Gardens 31
Bronte Beach 1, 179
Campbell Street 30
Centennial Park 8
Chinatown 30
Combined High Schools 14
E.F.G. Bookshop 30
Great Public Schools 14
Hotel Australia 30, 180
Kensington 1
La Perouse 9
Lang Road, Paddington 7
Lidcombe 1
Maroubra 9
Moore Park 9
Moore Park Road 9
Oxford Street, Paddington 7
Paddington 7, 180
Poate Road, Paddington 6, 179
Point Piper 20
Royal Agricultural Showground 9
Roycroft Bookshop 30
Tamarama Beach 6, 180
Wellington Street 2
Sydney Boys' High School 9
Sydney Harbour 32
Sydney Hospital 37
Sydney Medical School 35
Sydney Public Library 26
Syme 71

Tadokoro, J. 159, 185
Tatum, E. L. 157
Taylor, Sherwood 11
The Times 162, 163, 192
These Ruins are Inhabited 112
thoracic duct 99
Thurber, James 156
Times Literary Supplement 222
Times Square 67
Tit Bits 193
Tjio, J. H. 199
Tolkien, J. R. R. 71
Tomkins, G. M. 125
Toynbee, Arnold 29
trams 8
transfer of nuclei 197
Trendall, A. D. 28

INDEX

Trudeau Institute for Medical Research 133
trypsin 88
tuberculosis 64
tumours, mouse 218
Turner, Peggy J. 65, 163, 165
Tyrwhitt, Ursula 91

US Customs 118
Ulysses 30, 132
United States of America 25, 111, 117
University of Cambridge 147
  Chair of Pathology 151
  Department of Pathology 152
University of Chicago 135
University of East Anglia 160
University of Melbourne 42, 49
  Physiology Department 49
University of Oxford
  Australian professors 72
  bacteriology 69
  Chair of Pathology 162, 175
  Church of England 72
  clinical professors 227
  clinical school 227
  college chaplains 73
  college life 176
  college teaching 175
  colleges 177
  Congregation 174
  D.Phil. 65, 80
  D.Phil. students 203
  Dean of Degrees 65
  Department of Physiology 166
  Encaenia 72
  Faculty Board, Clinical Medicine 226
  Faculty Board, Physiological Sciences 226
  Faculty Boards 175
  Final Honour School of Animal Physiology 58, 69
  General Board of the Faculties 174, 226
  Goldsmith's Professor of English 132
  gowns 72
  Halifax House 71
  headship of colleges 232
  Hebdomadal Council 73, 174, 226
  in statu pupillari 66, 80
  Keble Triangle 76
  Magdalen College 163
  matriculation ceremony 66
  medical students 69
  Newcomers' Club 129
  Oxford University Gazette 174
  pathology 69
  post-graduate instruction 203
  Professorship of Pathology 208, 225
  Queen's College, Provost's Lodgings 166
  religious observances 72
  research 167
  Rhodes House 55
  science 71
  science departments 55
  Sheldonian Theatre 65
  South Parks Road 55, 71, 171
  subfusc 80
  Theology Faculty 73
  tutorial fellows 68
  tutorial system of teaching 175, 176
  undergraduate lectures 171
  University Church 73
  University Parks 164
  Vice-Chancellor 174, 224, 225
University of Sydney 19, 23, 73
  Department of Medicine 40
  Faculty of Arts 23
  Faculty of Law 23
  medical course 34
  New Medical School 36
  Old Medical School 36
University of Washington, Seattle 202

Van Heyningen, W. E. 64
Vendée 184
Vermont, autumn in 130
Vicia faba 151
Victoria, Australia 48
  National Gallery of 49
Vienna Circle 29
Villejuif 191
virus, inactivated 188
viruses 160
Vogel, H. J. 182

Wallace, R. A. 34
Walt Disney 192

war effort 24
waratah 19
Warburg, O. 92, 95, 198, 211, 212
Ward, H. K. 38, 44
Washington, D.C. 97, 119, 122, 137
  Massachusetts Avenue 122
  Potomac River 130
  Rock Creek Park 130
Waterhouse, E. G. 26
Watkins, J. F. 185, 192, 194
Watson, G. M. 63, 86, 98
Watson, J. D. 76
Watt, Fiona M. 201
Watts, J. W. 103, 106, 123, 137, 139, 140
Weiss, Mary C. 201
Wellcome Trust 174
Wells, A. Q. 63
Welt im Wandel 193
White, J. 119
'White Australia' policy 127
white cells of the blood 61
Who's Who 207
Wiener, F. 219
Wilhelmshaven 198
William De La Pole, Earl of Suffolk 228

Williams, T. I. 75, 100
Williamsburg, Virginia 130
Willis, R. A. 82
Wittgenstein, Ludwig 29
Wöhler,, Friedrich 10, 187, 191
Woodin, A. M. 166
World of Science 11
Wright, R. D. 42, 101
Wylie, J. A. H. 64

X-rays 202

Yanofsky, C. 135
Yellow Book 91
Yerganian, G. 198
Young, F. G. 152

Zamecnik, P. C. 133, 134
Zech, Lore 220
Zinsser, Hans 38
zoology 34